Fibre structure

Fibre structure

Dr Siba Prasad Mishra

WOODHEAD PUBLISHING INDIA PVT LTD

New Delhi, India

Published by Woodhead Publishing India Pvt. Ltd.
Woodhead Publishing India Pvt. Ltd., 303, Vardaan House, 7/28, Ansari Road, Daryaganj, New Delhi - 110002, India
www.woodheadpublishingindia.com

First published 2016, Woodhead Publishing India Pvt. Ltd.
© Woodhead Publishing India Pvt. Ltd., 2016

Reprinted 2020

Woodhead Publishing India Pvt. Ltd. ISBN: 978-9-38505-913-1
Woodhead Publishing Ltd. e-ISBN: 978-9-38505-963-6

Typeset by Third EyeQ Technologies Pvt Ltd, New Delhi
Printed and bound by Replika Press Pvt. Ltd.

Contents

Contents

Preface

Fibre Structure is an important as well as fundamental subject for the background knowledge and information for textile science, fibre science, chemical processing as well as in technical textiles. Although it is fundamental in nature, still then it covers lot of information related to the structure, like type of structure, arrangement of different components of structure and over all its visualization. Equally, there are so many different instruments with their own unique principle for investigation of the structure. In addition, there have been great advances in the subject, and the whole field of fibre structure is too large to cover coherently.

In spite of all these developments and diversifications in structural measurement and structural analysis, there is no specific book for first hand information. And so there is a need for a Fibre Structure book for the basic level, with a proper understanding.

With this concept and background, I have made my attempt to write this book in a compact and simplified manner to cover the initial information of the subject. Some of the materials presented in this book were experimented by me as part of my project as well as research activity. I hope this book will be helpful for all the students, researchers, analysts as a stepping stone to understand the related subjects.

I am thankful to all who always encourages me to write and communicate to all, whatever knowledge I gathered during my student life and professional life. Above all, I am equally thankful to Shilpi for her encouragement on her own way and own style. My special thanks owes to Woodhead Publications to come forward spontaneously for the publication of the book.

Dr. S. P. Mishra

Introduction

1.1 Introduction

Application of fibres in textiles and clothing starts from the development of human evolution. Clothing became existence, not for aesthetic purpose or decoration but for protection against cold, hot, rain, dust, etc. initially. Application of fibres in clothing dates back to 5000–4500 BC with utilization of hemp, flax, cotton, silk and later sericulture. Till Industrial Revolution, all the source of fibre is from nature. 18th and 19th centuries witnessed an era of industrial revolution along with machineries for fibre processing and application. Gradually with the introduction of regenerated fibres like rayon and later on synthetic fibres like nylon and polyester reduced monopoly of natural fibres. It also created a competition for new and unique fibres and fibre development.

The concept of regenerated and synthetic fibres induced a new dimension to the research activity to develop new fibres. It created a strong philosophy that from any material fibre can be developed, manufactured or regenerated provided the chemical behaviour and structure, the structural alignment, molecular properties and the processing conditions can be analysed and controlled. This means that there will be a wide range of fibres and equally a wide range of materials to be developed as fibres.

1.2 Classification of textile fibres based on sources

With this concept, the classification of fibre was established as per its source and it is mentioned in Fig. 1.1. The major natural fibres present around us from vegetable and natural sources. There are around 15 important natural fibres available for processing and conversion into fabrics. Those are discussed below.

1.2.1 Vegetable sources

Major fibres from vegetable sources are discussed below:

Cotton is most widely used natural fibre and consists of pure cellulose. It is produced in China, Brazil, India, Pakistan, USA and Uzbekistan.

Flax is a lignocellulosic bast fibre, mostly present in European Union. This fibre is mostly used to make linen.

Hemp is also a lignocellulosic bast fibre with low quantity of lignin. The world's leading producer of hemp fibre is China.

Jute is the strongest vegetable fibre from India and Bangladesh. It is also a lignocellulosic fibre.

Ramie is also a lignocellulosic bast fibre mostly available in China and Brazil. It is also known as China grass, with a silky lustre and better elasticity.

Sisal is a hard and coarser leaf fibre, mostly available in Brazil, Tanzania and Kenya.

Abaca is a leaf fibre, also known as manila hemp, extracted from leaf sheath around the trunk of Musa textiles. The world's major fibre producer is Philippines. Lignin content in the fibre is about 15%.

Coir is a hard, short and coarse fibre extracted from the shells of coconut. It is mostly present in India, Sri Lanka, Philippines, Vietnam, Indonesia and Brazil. This fibre contains highest amount of lignin making it stronger but less flexible.

1.2.2 Animal sources

Major fibres from animal sources are discussed below:

Alpaca is a hair fibre like wool, comes from the Lama Pocos. This fibre comes in approximately 22 natural colours, produced mostly in Peru, North America, Australia and New Zealand. It is stronger than wool fibre.

Angora is a rabbit fibre, very soft, fine and silky. 90% of the fibre is produced in China. Angora fabric is very suitable for thermal clothing.

Camel hair is available from the two humped Bactrian camel mostly present with nomadic households in Mongolia and inner Mongolia, China. It is the softest and more premium hair fibre.

Cashmere fibre is available with Kashmir goats, in China, Australia, India, Pakistan, New Zealand, Turkey and USA. It is a luxurious and expensive fibre.

Mohair fibre is produced from Angora goat, available in South Africa. It is a smooth and lustrous fibre.

Silk is the natural filament fibre, with high lustre, mostly produced in China, Brazil, India, Thailand and Vietnam.

Wool is the most important protein fibre. It is the first domesticated fibre, mostly produced in Australia, New Zealand, China, Iran, Argentina and UK.

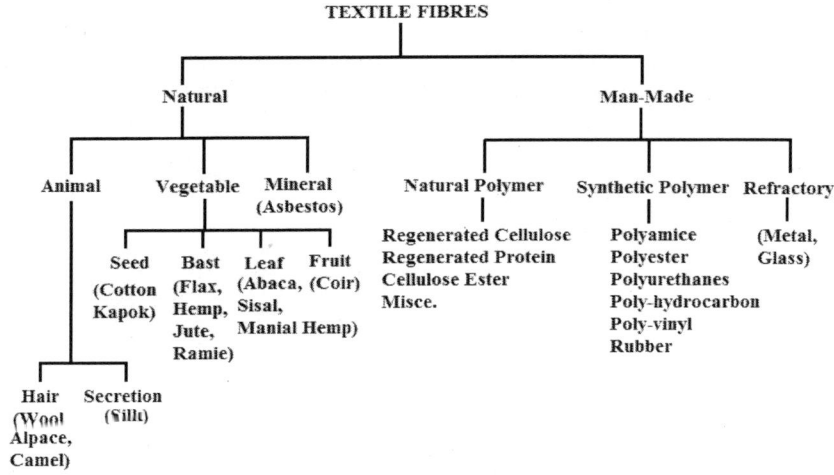

Figure 1.1   Classification of textile fibres as per sources.

1.2.3 Ground and petrochemical sources

In addition to the collection of the fibres from the sources above the ground, there are fibres from below the ground like metals. From World War II, there has been a thrust to produce synthetic materials, mostly derived from petrochemicals. The manufactured fibre is termed as 'synthetic fibres' as the raw materials were available by synthesis followed by polymerization and fibre formation. Synthetic fibres became the consequence of spectacular growth in petrochemicals development and utilization. The growth in the development of synthetic fibres and synthetic fibre industry along with polymer industry became phenomenal with the growth of petrochemical industry.

1.3   Classification of textile fibres based on polymer

Polymer is a material constructed of smaller molecules of the same substance that form larger molecules. The polymers are any of numerous natural and synthetic compounds of usually high molecular weight and consisting of up to millions of repeated linked units, each a relatively light and simple molecule. The term is derived from the Greek words: 'polumeres', where polus meaning many, and meros meaning parts. A key feature that distinguishes polymers from other molecules is the repetition of many identical, similar or complementary molecular subunits in these chains.

Polymers, macromolecules, high polymers and giant molecules are basically same and consist of high-molecular-weight materials composed of these repeating subunits. These materials may be organic, inorganic or organometallic, and synthetic or natural in origin. Polymers are essential materials for almost every industry such as adhesives, building materials, paper, cloths, fibres, coatings, plastics, ceramics, concretes, liquid crystals, photo resists and coatings.

These polymers can be natural or synthetic and organic or inorganic. Organic polymers are distinguished from inorganic polymers because of presence of carbon atom in the main chain. Presence of totally carbon atoms termed as carbochain polymers. If the main chain consists of other atoms with carbon, then it is termed as heterochain polymers. Natural inorganic polymers include sand, asbestos, agates, feldspars, mica, quartz and talc. Natural organic polymers include polysaccharides or polycarbohydrates such as starch and cellulose, nucleic acids, lignin, rubber and proteins. Synthetic inorganic polymers include boron nitride, concrete, many high-temperature superconductors and a number of glasses. Synthetic organic polymers include fibres, plastics and coatings, such as polyethylene, polypropylene, polyamides, polyesters, vinyl polymers, polyurethanes and synthetic rubbers. Fibres are polymeric materials that are strong in one direction, and they are much longer (>100 times) than their width. This is termed as l/d ratio. Elastomers or rubbers are polymeric materials that can be distorted through the application of force, and when the force is removed, the material returns to its original shape. Plastics are materials that have properties between fibres and elastomers—they are hard and flexible.

The resources for natural fibres are also natural high molecular weight polymeric substances. This means that both natural and synthetic fibres are polymeric materials. Based on the polymeric materials present in fibres, all fibres can also be classified in the way, shown in Fig. 1.2.

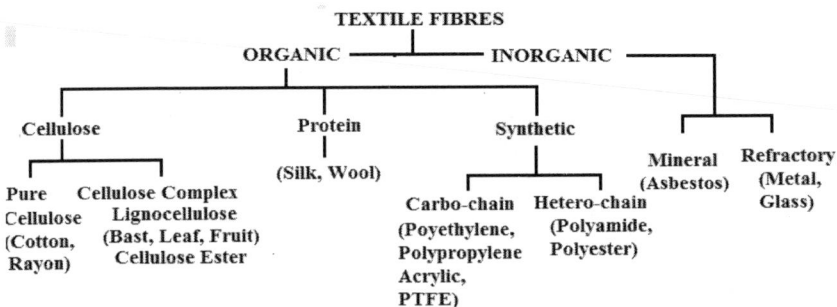

Figure 1.2   Classification of textile fibres based on the polymer.

1.4 Concept of fibre science

Fibre science as well as polymer science is related to the physics and chemistry of large chain molecules. Long before the complexity of the polymer was understood, the chemists were taken first step to copy their molecular structure. The theory of polymer structure was advocated by Staudinger from 1920. He put forward the theory that polymers are built up from chain molecules. The chain molecules can also lead to a three-dimensional network polymer. Carothers and his co-workers began their work on polycondensation. He put forward that polymers of definite structure could be obtained by the use of classical organic reactions. The properties of the polymer can be controlled by the starting compounds, i.e. monomers. Although his work was based on research in condensation polymers, however, his theory also hold good for addition polymers.

The polymer molecules may be long and straight chains and they may be branched with small chains extending out from molecular backbone. The branches may grow until they entangle with other branches to form a huge, three-dimensional matrix. The molecular chains can be chemically interconnected so that they will give cross-linked polymers. The geometric structure of the polymer can be expressed by configuration and conformation. Configuration refers to the order that is determined by chemical bonds and cannot be altered by physical means. Conformation refers to the order that arises from the rotation of molecules about single bond. The physical properties as well the processing conditions for the formation of fibres are determined by the configuration of the constituent atoms and also to the molecular weight. The polymeric systems differ from other materials like metals, ceramics and even biological systems in their disorder. This leads to different types of structures like crystalline, fibrils, micelles or spherulitic or multiphase systems.

The science and technology of the fibre is influenced by its chemical and molecular theory and the respective structure. All fibre-forming materials are basically polymers. For a linear polymer chain, the following factors are important for their suitability as fibres. Those are:

1. Intermolecular features of flexibility and internal resistance of the chain to change of shape, which will resist its collapse from its elongated condition to a randomly coiled configuration.

2. Intermolecular forces which will sustain side by side placement of molecules imposed by orientation.

3. Shape and symmetry factors which will assist the close packing of the chain and lead to crystalline structure.

1.5 Concept of fibre structure

This shows complex morphological systems being composed of different phases with variations in molecular weight, tacticity, chemical composition, as well as molecular and morphological structure. The utility of any fibre is dependent upon the structure of the fibre formed during formation and induced during post-formation processes. The major theoretical problem is the relation between the structure and the resultant properties. Hence, it is important to study different structures and their importance in fibre properties and thus application. The utilization of any materials like fibres and thus their development is largely dependent upon: (a) their mechanical properties under end use conditions, (b) mechanical properties under processing conditions and (c) interaction between the influence of processing variables upon structure and performance of the product. The performance of the fibres due to loading depends upon the molecular structure and the spatial arrangement of the molecules. The molecular structure of the fibre is generally established during polymerization. All these require a complete understanding of the fibre structure. Structural investigations should lead to the analysis of chemical structure, tacticity or configuration of the polymer, chain conformation and crystallinity. Structural investigation as well as the information available is not simple and straightforward. It is required to investigate a wide range of information from several techniques. Each technique has its own science and theory. Some of this information might be circumstantial and qualitative and might be included artefacts and errors in experimental techniques and calculations.

The structure of any polymer or fibre is generally present in the following way:

(1)   A number of small molecules joined together to form a long chain molecule, like that of a chain link.

(2)   The atoms are linked in the molecule by means of covalent bonds.

(3)   Long chain molecules are arranged together.

(4)   Some molecular chains lie parallel and sufficiently close together to act as a unit, known as crystallite.

(5)   Some molecular chains are arranged at random to form amorphous region.

(6)   Some long molecular chains run through the amorphous region and also two or more crystallites to form a network, which holds the structure together.

To obtain complete understanding of the fibre structure, it is desirable to obtain

(a) The chemical constitution of the fibre, i.e. presence of different atomic groups and the connecting chemical bonds.

(b) The configuration of the individual polymer chains, i.e. the spatial arrangement of the atoms within the molecule.

(c) The arrangement or the conformation of the chain molecules.

(d) The arrangement and nature of inter chain bonds.

(e) The arrangement of a complete three-dimensional molecular arrangement.

(f) Relationship of the chain molecules within itself and with one another.

(g) The arrangement of the molecules to form crystallites, i.e. side by side geometrical arrangement of the chains.

(h) The relationship of the crystal lattice to the overall structure of the fibre.

(i) The arrangement of the crystallites in the macroscopic body.

The analysis of the fibre structure or the polymer structure can be evaluated by means of the following steps:

- Chemical structure of the fibre
- Molecular weight
- Length of the chain molecule
- Crystal structure of the fibre
- Arrangement of the chain molecules
- Arrangement of the intermolecular forces
- Dimensions of crystallites
- Nature of the non-crystalline regions separating the crystallites with one another
- Crystallites and crystallinity relationship
- Arrangement of the crystallites in the fibre structure

The quantitative information may also be dependent upon certain assumptions and presumptions. Further, the theories are not mathematical and may not be formulated accurately. The structural ideas may change with change in information collected from different techniques with different theories. Hence, it is essential to have first hand information about the structure and different techniques used for structural information.

The chemical structure or the constitution of the atoms or the atomic groups is in the order of 1–4 Å and the spacing between the chain molecules is within the range of 4–10 Å. The intermolecular forces can be effective when the molecular chains are within the range of 2–7 Å. A single chain molecule will have a molecular weight of 20,000 or more and the length will be at least 500 Å. The unit structural element of the polymer or the fibre is in the range of 6–30 Å. The dimension of the crystallite is in the range of 50–200 Å, where the thickness/diameter will be less than that of the length in case of oriented system. The spherulites if present in the structure will have a dimension of around 5000 Å–100 µ. The diameter of the fibre is less than 1 mm. The measurement of these complicated structures cannot be completed by means of a single technique. Different techniques with different principles are used to evaluate different structural phenomenon. The techniques are based on the following theories.

(1) **Microscopy methods**: These methods are used to study the fine structure and morphology of the polymer or the fibre by means of a microscope and with aid of light or electrons. Accordingly, there will be light microscopes or electron microscopes. The light microscopes can any of the types like compound microscopes, polarizing microscopes, phase contrast microscopes and interference microscopes. The electron microscopes can be scanning electron microscope (SEM) or transmission electron microscope (TEM), depending upon the movement of the electrons.

(2) **Diffraction methods**: These methods are used to study the structure, particularly the crystalline structure of the polymer or the fibre in any diffractometer with aid of x-ray or electrons, i.e. wide angle x-ray diffraction (WAXD) or electron diffraction (ED). The instrument is known as diffractometer.

(3) **Scattering methods**: These methods are used to study the microscopy structure of the polymer or the fibre from the scattered rays. The scattered rays may be light, neutron or x-ray. The diffraction angle in scattering is always less than 2° for which the phenomenon is known as small angle x-ray scattering (SAXS), small angle light scattering (SALS) or small angle neutron scattering (SANS). Diffraction of scattering at small angles is associated with the colloidal to mono scales and it is the size range of a typical polymer chains. The colloidal scale is also associated with polymer crystallites or lamellae.

(4) **Spectroscopic methods**: Spectroscopy is used to study the chemical composition and molecular topology. These methods are used to study the structure of the polymer or the fibre by interaction with electromagnetic radiation at a continuous wavelength. The information

obtained by these methods, which are collected at different wavelengths is known as spectrum. The instrument is referred as spectrophotometer and the phenomenon as spectroscopy. The spectroscopy may be electronic spectroscopy, vibrational spectroscopy or resonance spectroscopy. Different methods in electronic spectroscopy are electronic spectroscopy for chemical analysis (ESCA), Ultraviolet (UV) spectroscopy or visible spectroscopy. Infrared spectroscopy and Raman spectroscopy comes under vibrational spectroscopy. Nuclear magnetic resonance (NMR) or electronic spin resonance (ESR) comes under resonance spectroscopy.

(5) **Thermal methods**: This method is done by heating the sample under controlled atmosphere and its effect on specific properties of the polymer or the fibre like temperature, specific heat or weight will highlight information about the polymer or the fibre. Thermal analysis is useful in describing solid state transitions in polymers and fibres. This also helps to understand the mechanical properties of the fibres as well as on processing.

The information on different structural parameters and their probable techniques for investigation is shown in Table 1.1.

Table 1.1   Different techniques for structural analysis

Structural parameters	Investigating techniques
Spherulites	Optical microscopy (OM)
	Electron microscopy (EM)
	Electronic spectroscopy (ESCA)
	Scanning electron microscopy (SEM)
Lamellar length	Small angle light scattering (SALS)
	Transmission electron microscopy (TEM)
Crystallite length	Small angle x-ray scattering (SAXS)
Lamellar width	Thermal methods
Crystallite thickness	Wide angle x-ray diffraction (WAXD)
Crystal structure	Electron diffraction (ED)
Chain folding	Transmission electron microscopy (TEM)
	Small angle x-ray scattering (SAXS)
	Raman spectroscopy (R)
Chain separation	Small angle neutron scattering (SANS)
Tacticity	Nuclear magnetic resonance (NMR)

Constitution	Raman spectroscopy (R)
	Infrared spectroscopy (IR)
	Nuclear magnetic resonance (NMR)
	Electronic spin resonance (ESR)

Further readings

1. S. P. Mishra, *A Text Book of Fibre Science and Technology*, New Age International Publications, New Delhi, 2000.

2. M. Lewin and E. M. Pearce (Eds.), *Handbook of Fibre Science & Technology*, Marcel Dekker, New York, 1985.

3. Frederick T. Wallenberger and Norman E. Weston, *Natural Fibers, Polymers and Composites*. Kluwer Academic Publishers, Boston, 2004.

2.1 Introduction

Chemical structure is the arrangement of chemical bonds between atoms in a molecule. It is basically the relation between atoms and neighbouring atoms to form the molecule along with the geometrical shape of the molecules. For macromolecules like fibres and polymers, chemical structure is the molecular geometry. Molecular geometry refers to spatial arrangement of atoms that hold the molecule together. The adoption of definite chemical structures for polymers results in its practical applications, because it has led to an understanding of how and why the physical and chemical properties of polymers or fibres change with the nature of chemical groups and structure.

2.2 Raw materials

Polymers are the basic material for all fibres: natural, man-made or synthetic. The respective polymer in the natural fibres or fibres from natural resources (regenerated fibres) formed during its growth. On the other hand, the synthetic fibres are manufactured from their respective polymers, which in turn are produced by a process, referred as polymerization. Polymer or macromolecule is defined as 'A molecule of high relative molecular mass, the structure of which essentially comprises the multiple repetition of units derived, actually or conceptually, from molecules of low relative molecular mass.' The polymeric raw materials for different important fibres are mentioned in Table 2.1

Table 2.1   Fibres and the respective polymers

No.	Fibre	Polymer
1	Cotton	Cellulose
2	Rayon	Cellulose
3	Jute/flax/hemp/ramie	Lignocelluloses
4	Sisal/manila hemp/abaca	Lignocelluloses

5	Coir	Lignocelluloses
6	Silk (fibroin/sericin)	Polypeptide/protein
7	Wool (keratin)	Polypeptide/protein
8	Alpaca/cashmere/mohair	Polypeptide/protein
9	Camel	Polypeptide/protein
10	Angora	Polypeptide/protein
11	Casein/ardil/soyabean	Polypeptide/protein
12	Polyester/terylene/dacron	Poly(ethylene terephthalate)
13	Acrylic	Poly (acrylonitrile)
14	Nylon	Polyamide
15	Dyneema	Polyethylene
16	Polypropene	Polypropylene
17	Nomex	Aromatic polyamide
18	Kevlar	Aromatic polyamide
19	Spandex	Polyurethanes
20	Teflon	Polytetrafluroethylene

2.3 Polymer molecule

The polymer molecule presents in fibres have the following characteristics:

1. Polymer molecules can be very large with high molecular weight, and also referred as macromolecule.

2. Small, simple chemical units joined together linearly to form the polymer.

3. Small molecules from which polymer is formed is knows as a monomer.

4. The repeat unit in the polymer chain is known as 'mer'.

5. Most polymers consist of long and flexible chains with a string of C atoms as a backbone.

6. The carbon atoms connected together by covalent bonds.

7. Double bonds are possible in both chain and side bonds.

8. Carbon atoms present in the main chain can have side-bonding with H atoms or radicals.

Structure of few commercial fibre/polymer molecules are mentioned in Fig. 2.1.

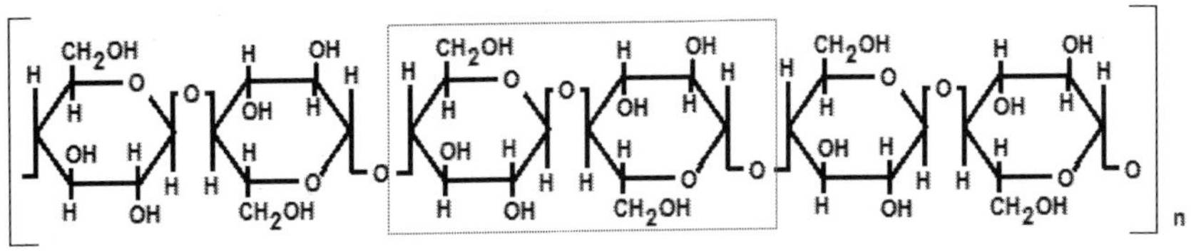

(a)   Cellulose

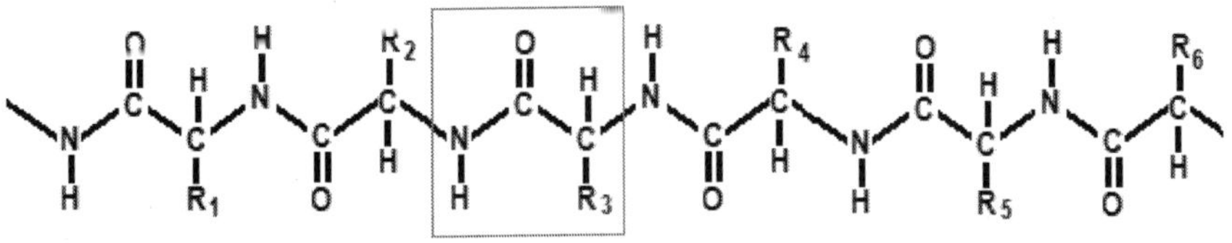

(b)   Protein

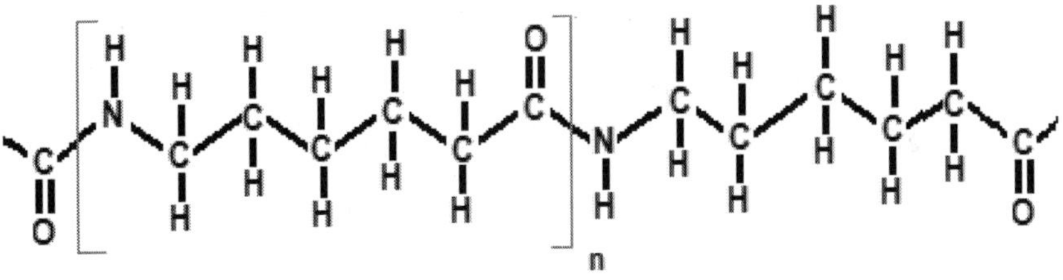

(c)   Polyamide 6

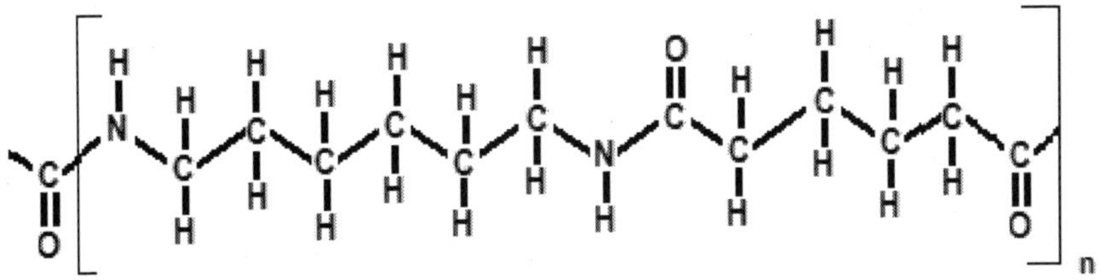

(d)   Polyamide 6,6

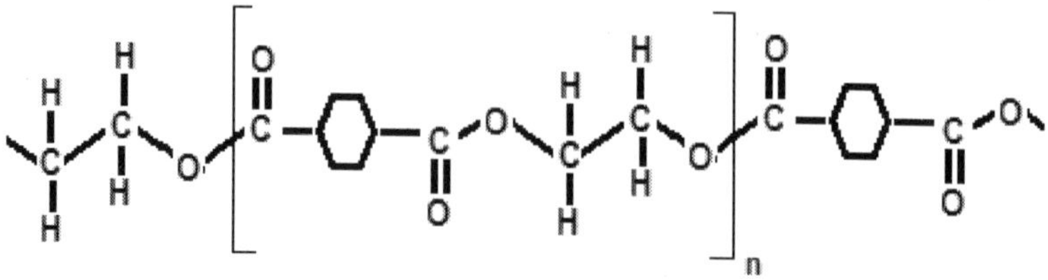

(e)   Poly (ethylene terephthalate)

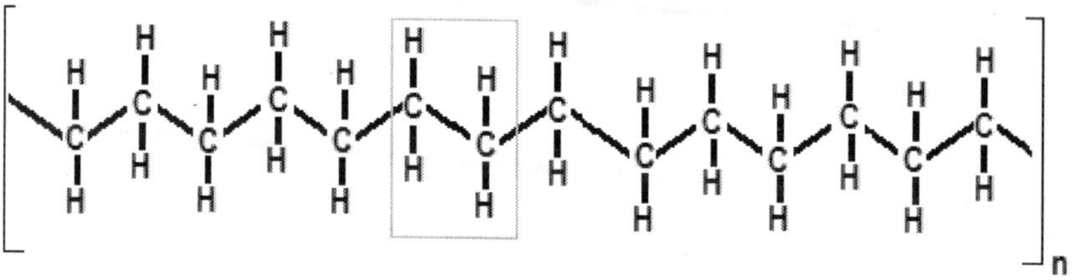

(f)   Polyethylene.

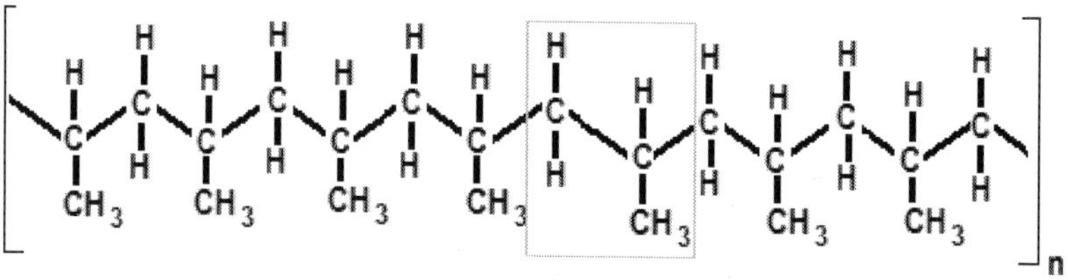

(g)   Polypropylene

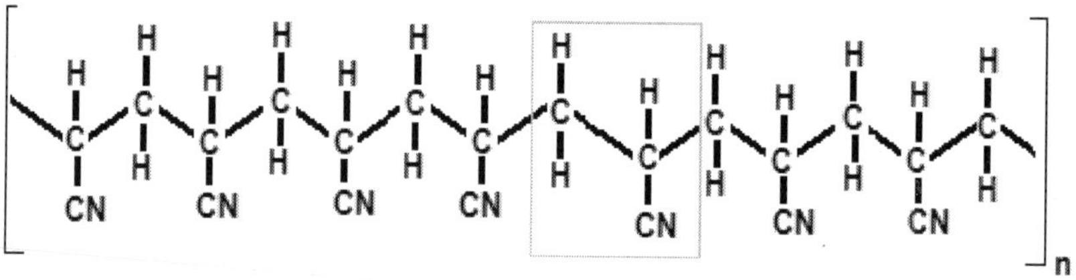

(h)   Polyacrylonitrile

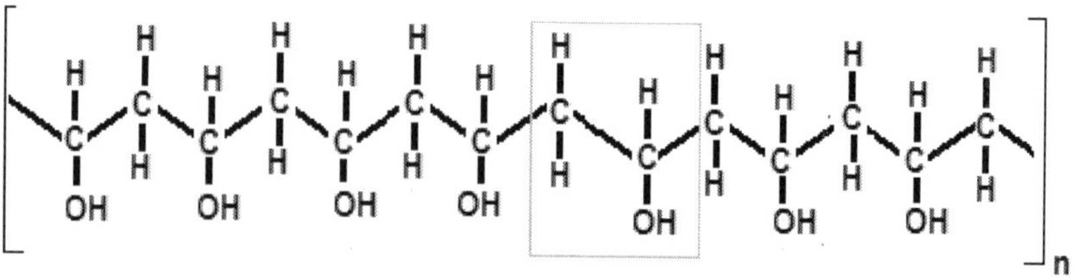

(i)   Polyvinyl alcohol

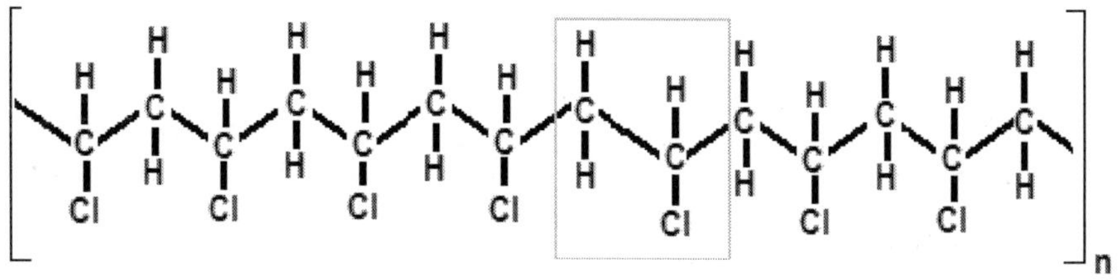

(j) Polyvinyl alcohol

Figure 2.1 Individual fibres/polymers molecule configuration with their repeat unit.

2.4 Chemical bonding in polymers

The molecular arrangement in polymers is related to definite atoms related to each other with chemical bonds and forms covalent bonding between neighbouring atoms. This bond is characterized by sharing of electron between two neighbouring atoms. This type of electron sharing keeps the shared electrons close to both atomic nuclei. One pair of shared electrons makes one covalent bond. A molecule is a group of atoms held together by covalent bonds. Two nonmetallic atoms can share valence electrons with each other. There are no elements willing to become cations, so ionic bonds are not possible.

To determine how many covalent bonds can be formed between atoms, first the number of valence electrons must be counted. One shared pair of electrons is a single bond. Two shared pairs (four electrons) make a double bond, and three shared pairs (six electrons) make a triple bond. The more electrons that are shared, the stronger the bond will be. The neighbouring elements, namely, carbon, nitrogen and oxygen commonly use double and triple bonds. Nitrogen needs a triple bond to achieve octets for each atom, but a double bond is sufficient for the oxygen molecule. The type of covalent bond affects the shape of a molecule. The nuclei move closer together if they share more electrons. This means that a triple bond is shorter than a double bond, which is shorter than a single bond. The bond angles are also very specific in a covalently bond molecule. The shared electrons want to be near the two positively charged nuclei, but try to stay away from negatively charged lone pairs. This structure is quite different than that created by metallic bonds, which do not have a particular orientation.

Carbon is tetravalent means it makes four bonds to other atoms with four electrons available. Carbon can share with maximum four bonds and so can form single bond with four elements or double bond with two elements. Because it forms four covalent bonds, carbon displays tetravalency in the

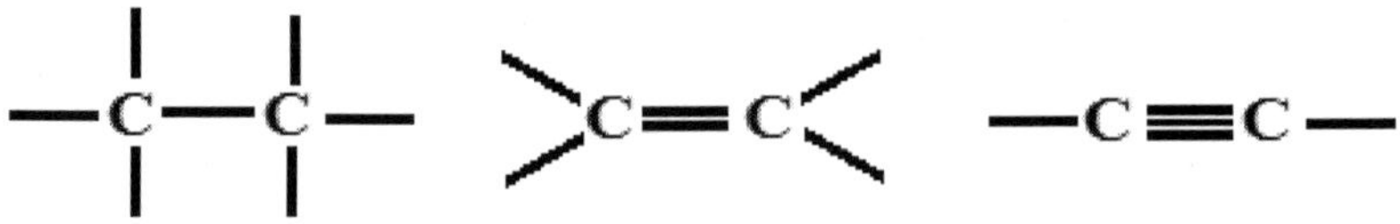

Figure 2.2   Bonding in carbon atom.

combined state, covalently bonded to one, two, three or four carbon atoms or atoms of different elements or group of atoms, as shown in Fig. 2.2

Carbon exhibits unique catenation property by which carbon atoms are linked through a covalent bond to various kinds of chains of carbon atoms. With the advantage of tetravalency and catenation, carbon forms large number of compounds formed by carbon alone exceeds number of compounds formed by all other elements in the periodic table.

2.5 Chemical structure

These macromolecular substances have the general formula $X - [A]_n - Y$, in which A is the repeating unit and X, Y are end groups. A is referred as 'mer' and it is a grouping of atoms that is different for different polymers or fibres and it may be simple like $- CH_2 -$ or even more complex. The polymers are formed by the process of polymerization from the respective monomers. A monomer is defined as 'A molecule which can undergo polymerization, thereby contributing constitutional units to the essential structure of a macromolecule.' Some important polymers and their monomers are listed in Table 2.2. The chemical structure of these polymers is shown in Fig. 2.2.

Table 2.2   Polymer and the respective monomer

Polymer	Monomer
Polyethylene	Ethylene
Polypropylene	Propylene
Poly (vinyl chloride)	Vinyl chloride
Poly (vinyl acetate)	Vinyl acetate
Poly (vinyl alcohol)	Vinyl acetate
Poly (acrylonitrile)	Acrylonitrile
Polyamide 6 (Nylon 6)	Caprolactam
Polyamide 66 (Nylon 66)	Adipic acid, Hexamethylene diamine
Poly(ethylene terephthalate) (Polyester)	Ethylene glycol, Terephthalic acid or Dimethyl terephthalate

Polymers are formed from the respective monomers by means of chemical reactions and the process is known as polymerization. These polymers can be formed by joining the monomers together with covalent bonds during polymerization process. The length of the polymer chain is specified by the number of repeat units (n) in the chain, often noted as degree of polymerization (DP). The value of n is usually large, more than 100 or more depending upon the product. The molecular weight is the product of the molecular weight of the repeat unit and the DP. The primary valency bonds are basically covalent bonds. The two participating atoms each contribute one electron to the common linkage and the two electrons are thus shared to unite the atoms with a bond of high energy between about 20 and 200 kcal/mol. This high energy makes them difficult to disrupt.

2.6   Classification of polymers

The polymerization process proceeds with many variations. Because of these variations, polymers can be classified in various ways as shown in Table 2.3.

Table 2.3   Classification of polymers

	Monomer characteristics	Types of polymers
1	Structure of monomers	Carbochain or heterochain
2	No. of monomers used	Homopolymer or copolymer
3	Chemical structure of the monomer	Addition polymers or condensation polymers
4	Functionality of the monomer	Linear polymers, branched polymers, cross-linked polymers
5	Chemical groups present in the monomer	Polymer named as per the chemical group, e.g. polyamide, polyethylene

(1)   Structure of monomers used

Polymers basically consist of monomers or repeat units, i.e. mer. The monomer consists of mostly carbon atom and few other atoms like nitrogen, oxygen, hydrogen. These other atoms may be present in the main chain or in the side chain. Depending upon the presence of carbon atom with or without any other atoms are either (a) carbochain polymers or (b) heterochain polymers. Carbochain polymers contain only carbon atoms in their main chain. Polyethylene, polypropylene, polyacetylene and all vinyl polymers are carbochain polymers. Heterochain polymers consist of other atoms also along with carbon in their main chain. Polyamides, polyesters, polyurethanes are heterochain polymers. Some polymers like polyacetylenes, polyester, etc. contain unsaturated bonds in their main chain.

(2)   Monomers used

Polymers like polyethylene containing a single repeating unit, such as ethylene, are called homopolymers. Homopolymers are defined as 'A polymer derived from one species of monomer.' So when a single monomer is polymerized into a macromolecule, it forms a homopolymers and so its mer unit is same. Polymers containing two or more different groups or monomers or structural units are called copolymers and the chain molecules contain different mer or different structural units.

There will be either one monomer or more than one monomers used for the formation of the polymer. Sometimes for enhancement of specific properties, comonomers are used and the polymers formed these monomers and comonomers are known as copolymers. A polymer derived from more than one species of monomer. Copolymers that are obtained by copolymerization of two monomer species are sometimes termed bipolymers, those obtained from three monomers terpolymers, those obtained from four monomers quaterpolymers, etc. Formation of the copolymer changes the chemical structure of the homopolymer and it leads to modification of the molecular structure as well as properties. Since a copolymer consists of at least two types of mer or monomers, copolymers can be classified based on how these units are arranged along the chain. These are shown in Fig. 2.3 and mentioned below:

- **Alternating copolymer** is 'A copolymer consisting of macromolecules comprising two species of monomeric units in alternating sequence,' as shown with regular alternating A and B units in Fig. 2.3A.

- **Periodic copolymer** is 'A copolymer consisting of macromolecules comprising more than two species of monomeric units in regular sequence,' as shown with A and B units arranged in a repeating sequence (e.g. $(A\text{-}B\text{-}A\text{-}B\text{-}B\text{-}A\text{-}A\text{-}A\text{-}A\text{-}B\text{-}B\text{-}B)_n$.

- **Random copolymer** is 'A copolymer consisting of macromolecules comprising more than two species of monomeric units in random sequence,' with random sequences of monomers A and B (Fig. 2.3B).

- **Statistical copolymer** in which the ordering of the distinct monomers within the polymer sequence obeys known statistical rules. It is defined as 'A copolymer consisting of macromolecules in which the sequential distribution of the monomeric units obeys known statistical laws.' An example of a statistical copolymer is one consisting of macromolecules in which the sequential distribution of monomeric units follows Markovian statistics.

- **Block copolymer** in which each monomer unit repeats in blocks (Fig. 2.3C). It is defined as 'A macromolecule which is composed of blocks in linear sequence.' Block is a portion of a macromolecule, comprising many constitutional units, that has at least one feature which is not present in the adjacent portions.

- **Graft copolymer** in which the second monomer linked to the first monomer from side or as a branch. This is a branched copolymer (Fig. 2.3D) and defined as 'A macromolecule with one **or** more species of block connected to the main chain as side-chains, these side-chains having constitutional or configurational features that differ from those in the main chain.'

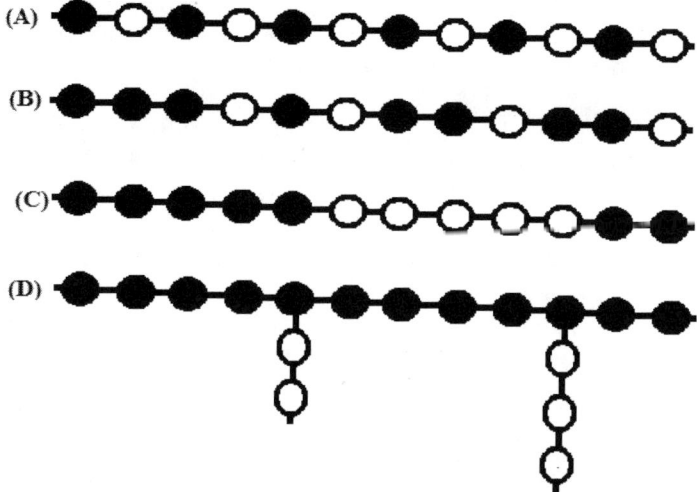

Figure 2.3   (A) Alternating copolymer, (B) Random copolymer, (C) Block copolymer and (D) Graft copolymer.

(3)   Types of monomers

Depending upon the structure or the types of the monomer used, the polymer can be classified as (i) Addition polymer and (ii) Condensation polymer.

(i)   Addition polymer

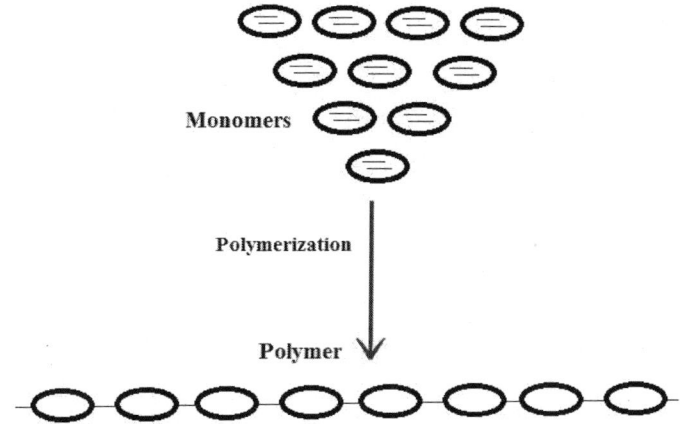

Figure 2.4   Formation of addition polymers.

The monomer used for addition polymer consists of double bond in its structure. The reaction starts when an activated site of a chemical is added to the double bond present in the polymer. It produces a new bond, a new activated location and has the capability to add up those monomers having double bonds. The process may be repeated thousands, or millions, of times to produce the macromolecule. An addition polymer is one in which the molecular formula of the repeating structural unit is identical to that of the monomer, e.g. polyethylene and polystyrene. The schematic reaction for the addition polymer formation is shown in Fig. 2.4.

(ii) Condensation polymer

The monomer used for this polymer, has at least two reactive groups, which reacts with one another of different monomers to form the polymer. Because of reaction, small molecular weight compounds will be eliminated during reaction. So a condensation polymer is one in which the repeating structural unit contains fewer atoms than that of the monomer or monomers because of the splitting off of water or some other substance. Polyesters and polyamides are the condensation polymers. The schematic reaction for the condensation polymer formation is shown in Fig. 2.5.

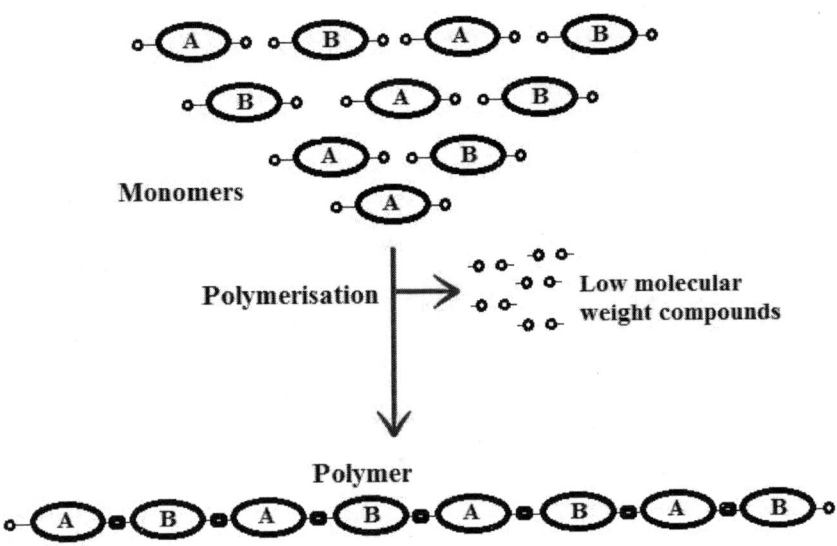

Figure 2.5   Formation of condensation polymers.

(4) Functionality

For the formation of the polymer, a monomer must have two reactant groups to be able to add to two molecules. This ability is generally known as 'functionality'. Functionality may be defined as the ability to form primary valency bonds. Depending upon the functionality, the polymer may be classified as (i) linear polymer, (ii) branched polymer, (iii) cross-linked polymer, and (iv) network polymer. The schematic structural diagram of a linear polymer, branched polymer, cross-linked polymer and network polymer is shown in Fig. 2.6. Functionality of 2.0 leads to the formation of linear polymers and more than 2.0 leads to the formation of branched or cross-linked polymer. In a linear polymer, the structural units are connected in a chain arrangement and thus need only be bifunctional, i.e. have two bonding sites. A trifunctional structural unit results in a nonlinear or branched polymer results. Ethylene, styrene and ethylene glycol are examples of bifunctional monomers, while glycerin and divinyl benzene are both polyfunctional. In branched or cross-linked polymer, the covalent bond is also present in a perpendicular manner to the main chain. In the former, it is not connected with the neighbouring chain, but in case of the latter, the covalent bond is connected from one neighbouring chain to the next neighbouring chain. This interconnection between two neighbouring chains by means of covalent bonds is termed as 'cross-linking'. When the density of the covalent bonds in parallel as well as in the perpendicular direction is almost same, then the cross-linked polymer is termed as network polymer.

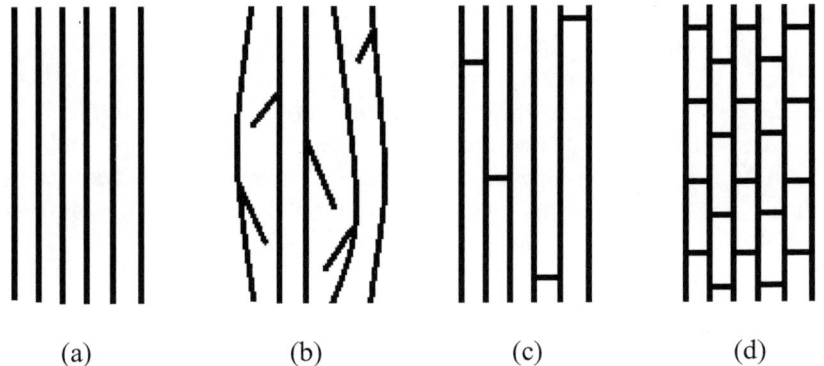

| (a) | (b) | (c) | (d) |

Figure 2.6   Structural representation of (a) linear polymer, (b) branched polymer, (c) cross-linked polymer and (d) network polymer.

2.7 Polymer structure phases

The structure of the polymer molecular chain in its simplest form consists of identical rigid segments connected by freely rotating joints. The structure of any polymer or fibre is more complicated and can be studied in three different phases.

1. **Primary structure**: This is also named as configuration. The sequence of repeat units within a polymer is called its primary structure. It

relates the exact order of arrangement of the atoms or molecules in the polymeric chains forming the long chain molecule.

2. **Secondary structure**: This is referred as conformation. Secondary structure refers to the localized shape of the polymer, which is often the consequence of hydrogen bonding. It also indicates the geometrical configuration of the long chain polymeric chain.

3. **Tertiary structure**: The tertiary structure refers to the overall shape of a polymer. This indicates the spatial arrangement of the molecular chains and the interactions between the side groups of the chain or chains.

4. **Quaternary structure**: This refers to the arrangement in space of two or more polymer subunits, often a grouping of tertiary structures. Many crystalline synthetic polymers form spherulites.

2.8 Configuration

Each and every polymer has a chemical constitution. It is present in a definite configuration. Configuration refers to the order that is determined by chemical bonds. This defines that the atoms are present strictly and definite spatial arrangement. Configuration is the arrangement of atoms, which cannot be altered except by breaking and reforming primary chemical bonds. For polymer chain, it is also known as chain configuration. Each individual chain molecule will have its fixed configuration, where each of the atoms and bonds is fixed. When two different chain molecules of the same chemical constitution have different configuration, then it is termed as 'Isomerism'. Configuration of different fibre forming molecules is shown in Fig. 2.1.

2.9 Isomerism

Isomerism is the phenomenon whereby certain compounds, with same molecular formula, exist in different forms owing to their different organization of atoms. Isomerism is the existence of two or more molecules that have same molecular formula, same number of atoms but a different arrangement within the molecule or it can be said different configuration.

The structure of the fibre forming polymers consists of complex spacing of C, H, O and/or N. As these atoms repeating so many times in a particular order, there is every possibility that the polymers show 'isomerism'. Isomerism is the phenomenon related to different compounds having same molecular formula. The compounds with the same molecular formula but different arrangements of atoms with different properties are called isomers or isomerides. Isomerism becomes an important structural feature as the complexity of the molecules increases and this is very important for long polymer chains. There are various types of isomerism for polymers and all are based on the nature of the carbon bond and the way carbon bonds are oriented in space. There are two types of isomerisms and those are: (a) structural isomerism and (b) stereoisomerism. Chain isomerism, position isomerism, functional isomerism, metamerism and tautomerism are all parts of structural isomerism because of the arrangement of atoms or groups of atoms in the molecule without any reference to spatial arrangement. When the compounds have identical structure but differ in their molecular configuration are known as stereoisomers and this phenomenon is known as stereoisomerism. There will be different types of stereoisomerism like optical isomerism and geometrical isomerism. Stereoisomerism phenomenon is more important for polymer and fibre structure.

(i)   Structural isomerism

Isomers that differ in connectivity are called structural or constitutional isomerism. This is mostly happened in saturated hydrocarbons and having at least three carbon atoms. Their physical properties differ slightly and also in melting point and boiling points. The number of isomers increases rapidly with the number of carbon atoms in the structure like two for butane (C_4), three for pentane (C_5) and five for hexane (C_6). This type of isomerism concept, when applied to polymers, leads branching, where instead of extension of a methyl group, a much larger molecule is extended from side.

In polyethylene, we have five types of structure and those are: (1) very low density polyethylene (VLDPE), (2) low density polyethylene (LDPE), (3) linear low density polyethylene (LLDPE), (4) medium density polyethylene (MDPE) and (5) high density polyethylene (HDPE). The schematic structural diagrams of these polyethylenes are shown in Fig. 2.7. Polyethylene is a polymer with the repeat unit [CH_2CH_2] and so it is expected to have a linear polymeric structure. But during polymerization and with specific condition, it gives rise to different types of polyethylenes like VLDPE, LDPE, LLDPE and HDPE. VLDPE is a substantially linear polymer with high levels of short-chain branches. LDPE is having branched with very long chains at infrequent points along the main chain. This is LDPE, having a small number of long chain branches. On the other hand, LLDPE is also a polyethylene having large number of short branches. HDPE is a highly linear chain without branching. Because of their differences in density, they are referred as low density and high density. This type of structural variation because of branching hinders crystallization and affects packing of chain molecules and thus results in different density. VLDPE is defined by a density range of 0.880–0.915 g/cm^3. LDPE is defined by a density range of 0.910–0.940 g/cm^3. LLDPE is defined by a density range of 0.915–0.925 g/cm^3. MDPE is defined by a density range of 0.926–0.940 g/cm^3. HDPE is defined by a density of greater or equal to 0.941 g/cm^3.

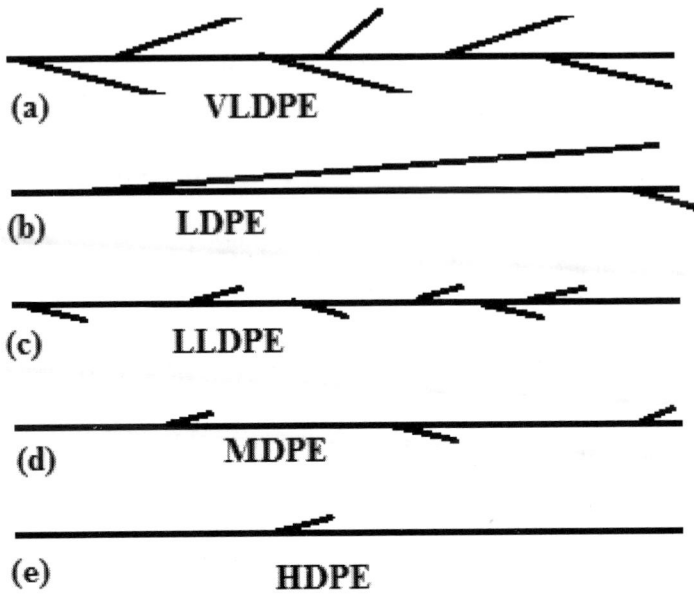

Figure 2.7 Schematic structural representation of (a) VLDPE, (b) LDPE,
              (c) LLDPE, (d) MDPE, and (e) HDPE.

(ii) Geometric isomerism

Geometric isomerism is because of the spatial arrangement of groups and
atoms. The locking into the spatial positions occurs because of a double bond
or a ring structure. The isomerism in which identical groups or atoms are
closer or same side is known as cis-form and the other, in which they are
remote is called trans-form. Both these words are Latin words where cis means
on this side and trans means across. Because of the restricted rotation around
the double bond, different isomers exist with particular groups on the same
side or on the opposite side of the double bond. The transition from cis to trans
configurations is impossible without the rupture of the chemical bonds, since
the rotation about the double bonds is impossible. This type of isomerism
occurs in butadiene and polyisoprene. The schematic structural formula is
shown in Fig. 2.8. In the figure two sides of the chain are in the same side
of the double bond and give rise to cis structure. In another figure, the two
parts of the chain are in the opposite sides of the double bond and it gives the
overall chain a zigzag appearance. In a similar way, cis–trans isomerism is
present in case of a ring structure. Cis and trans configuration is also present
in poly (ethylene terephthalate) due to the presence of C=O groups and shown
in Fig. 2.9.

(a) Cis   (b) Trans

Figure 2.8   Cis and trans configuration.

(a) Cis   (b) Trans

Figure 2.9   Cis and trans configuration in poly (ethylene terephthalate).

(iii)   **Stereoisomerism**

Stereoisomerism phenomenon or optical isomerism is related to the isomers, or different compounds having same molecular formula, same sequence of bonded atoms and same chemical properties but differs in three-dimensional orientations of atoms in space. This differs from structural isomerism, where there is difference in bond connections or order. Stereoisomerism is also referred as optical isomerism due to its optical property. The compounds are called stereoisomers or optical isomers. These isomers are said to be optically active because they rotate the plane-polarized light in either a clockwise or in a counterclockwise direction. The isomer that rotates the plane of plane-polarized light to the right (clockwise) is said to dextro-rotatary (d–); the other isomer will rotate the plane an equal amount to the left and said to be laevo-rotatary (l–). Optical activity of the organic compounds is dependent upon their molecular structure. All optically active organic compounds contain one or more carbon atom, whose four valences have been satisfied by four different monovalent groups or atoms. Such carbon atoms are called asymmetric carbon atoms. These compounds exist in different forms that are not superimposable and show optical activity.

2.10 Tacticity

The analysis of chemical structure of fibres or polymers indicates that fibres having the same chemical constitution may possess different structures. This is generally referred as tacticity. Tacticity of the polymers can be ob-

served in those polymers, which have asymmetric carbon atom like polypropylene and vinyl polymers. The tacticity of polypropylene is more famous
for the formation of different type of structures. These compounds of the
type $R–CH=CH_2$ contain asymmetric carbon atom because of its link with R
group. As a result of this, these atoms are likely to form d– and l– configurations. The two forms $CH_2=CH–R$ and $CH_2=CR–H$ can be distinguished
by arbitrarily assigning them as d– and l– configurations. These d– and l–
distributions decide the chain tacticity. In the polymer chain the d– and
l– configurations can form in any manner like (a) chaotically (mixed dl), (b)
similarly (ddd or lll) or (c) regularly (dl, dl, dl). So three different configurations can result for polypropylene as well as vinyl polymers ($CH_2=CHR$),
i.e. isotactic, syndiotactic and atactic, depending upon the arrangement of
the substituent group R. For isotactic configuration, the entire substituent
groups R lie above or below the plane of the main chain. Here the atoms
of the repeat unit will have same configuration. If the substituent groups
lie alternately above and below the plane, the configuration is called syndiotactic. In syndiotactic polymer, there will be regular alteration of asymmetric carbon atom in their molecules. A random sequence of the group is
known as atactic polymer. Here the d– and l– configurations alternate chaotically. Atactic polymers are more random and amorphous. Isotactic and
syndiotactic polymers are more ordered and hence crystalline. This means
that the tacticity, i.e. the degree of stereoregularity of a polymer can affect
its degree of crystallinity. Chemical constitution and configuration are the
primary structural concepts needed for understanding. All vinyl polymers
show tacticity. Table 2.4 shows the chemical structure or configuration of
atactic, isotactic and syndiotactic polypropylene.

Table 2.4   Tacticity of polypropylene

(A)   Isotactic polypropylene

$$\begin{array}{cccccccccccc} H & CH_3 & H & CH_3 & H & CH_3 & H & CH_3 & H & CH_3 & H & CH_3 \\ | & | & | & | & | & | & | & | & | & | & | & | \\ -C & -C & -C & -C & -C & -C & -C & -C & -C & -C & -C & -C- \\ | & | & | & | & | & | & | & | & | & | & | & | \\ H & H & H & H & H & H & H & H & H & H & H & H \end{array}$$

(B)   Syndiotactic polypropylene

$$\begin{array}{cccccccccccc} H & CH_3 & H & H & H & CH_3 & H & H & H & CH_3 & H & H \\ | & | & | & | & | & | & | & | & | & | & | & | \\ \cdot C & -C & -C & -C & -C & -C & -C & -C & -C & -C & -C & -C- \\ | & | & | & | & | & | & | & | & | & | & | & | \\ H & H & H & CH_3 & H & H & H & CH_3 & H & H & H & CH_3 \end{array}$$

(C) Atactic polypropylene

$$
\begin{array}{ccccccccccccc}
H & CH_3 & H & H & H & H & H & H & H & CH_3 & H & H \\
| & | & | & | & | & | & | & | & | & | & | & | \\
-C & -C & -C & -C & -C & -C & -C & -C & -C & -C & -C & -C- \\
| & | & | & | & | & | & | & | & | & | & | & | \\
H & H & H & CH_3 & H & CH_3 & H & CH_3 & H & H & H & CH_3
\end{array}
$$

The properties of each type of polypropylene are different. In isotactic and syndiotactic, the side group is present uniformly and so polypropylene chains can have uniform packing and so it can crystallize. On the other hand, atactic polypropylene chains are irregular and so have non-uniform packing. It leads to restrictions in crystallization. The carbon atom in a vinyl polymer to which attached a side group other than hydrogen is another example of asymmetric carbon atom. In vinyl polymers, like polypropylene, every alternate carbon atom is asymatic, as shown in Fig. 2.10.

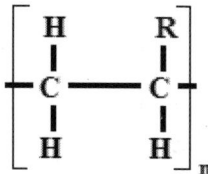

Figure 2.10   Vinyl polymer.

2.11   Rotational isomerism or conformations

Molecular conformation is basically related to the shape of the polymer molecular chain. The same polymer with same configurations but differs only by rotations about single bonds is said to be two different conformations of that polymer. The shapes are types of aggregation states of individual molecular chains. Conformation or rotational isomerism is the arrangement of the chain molecules, which can be altered by rotating groups of atoms around single bond. A single conformation is just a single shape that the chain can adopt. The molecular chain forms a zigzag arrangement due to the valence angle between the carbon atom and its two adjacent carbon atoms. The conformation in isolated state depends upon its chemical structure. The arrangement of the long chain molecules can be postulated in different ways as per any of the following manners like (a) Straight chain, fully extended, (b) Folded chain, coiled backward and forward or (c) Coiled chain, coiled into a spiral form. The schematic diagram of these structures is shown in Fig. 2.11.

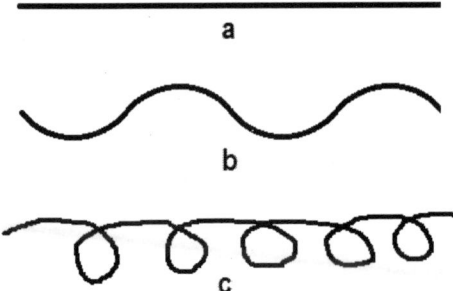

Figure 2.11   Representation (a) Straight chain, (b) Folded chain and (c)
Coiled chain.

Depending upon the solvents as well as the structure and chain flexibility, the conformation can vary from a straight chain to a coiled chain. The chain flexibility as well as the conformation influences melts flow properties, structure in the solid state, as well as the properties related to the molecular segmental mobility. This leads to the formation and development of the structure of the fibre or polymer during extrusion and post-extrusion and post-spinning operations. Natural fibres also exhibit different types of conformation. The examples of these types of chains in proteins are given at the end of this chapter. Straight chains are composed of long thread like molecules. Because of their minimum diameter, these chains can pack closely with high mutual attraction. So a stronger force will hold the neighbouring chains together. In case of helical chains, the molecules can be extended to their full length and more often the molecules folded or coiled into a long helix. When molecules are present in helical form, the chances of molecular association and entanglement taking place are very great.

The conformation, i.e. dimension of a chain is generally expressed in terms of any of the following two parameters. Those are: (1) mean square end-to-end distance and/or (b) mean square radius of gyration. The value of radius of gyration will lead to the evidences of different conformations of the polymer chain.

2.12   Model of a molecular chain

A polymer molecule is a big molecule composed of many small molecules chemically joined together. The molecular chains are flexible and the instantaneous shape of a polymer chain is referred as its conformation. Because of its flexibility, the molecule can be randomly coiled in the collapsed state and can be opened up progressively by stretching towards its limiting maximum extension. The unique property of the polymer or the fibre is its response to large deformation. Under external forces, the molecular chain becomes aligned in the direction of deformation. The most successful model relating to the deformation process to molecular fine structure is the proposed the theory of rubber like elasticity, consisting of a concentration of identical, rigid segments connected by freely rotating joints (Fig. 2.12). A single isolated polymer molecular chain can be assumed to have the characteristics like (a) Isolation of long chain, (b) Linearity, i.e. no branching, (c) Indefinitely long, (d) Small free volume and (e) Freely jointed links.

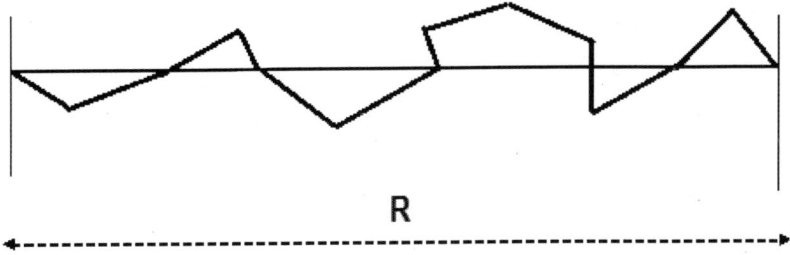

R

Figure 2.12   Model of a polymer chain.

The unit can rotate relative to its neighbour. The units can assume arbitrary orientation, i.e. any conformation. The model is substantially linear, not cross-linked or three-dimensional. The length is infinitely long, consists of 'n' rigid segments, each of length 'a'. Each segment is joined to its neighbour with and angle 'θ'. The average $\cos \theta$ for any polymer molecular chain is constant. One bond or one unit of the chain can rotate freely relative to its neighbour. Such a chain can assume any conformation and it is perfectly flexible, as shown in Fig. 2.13(a) and (b).

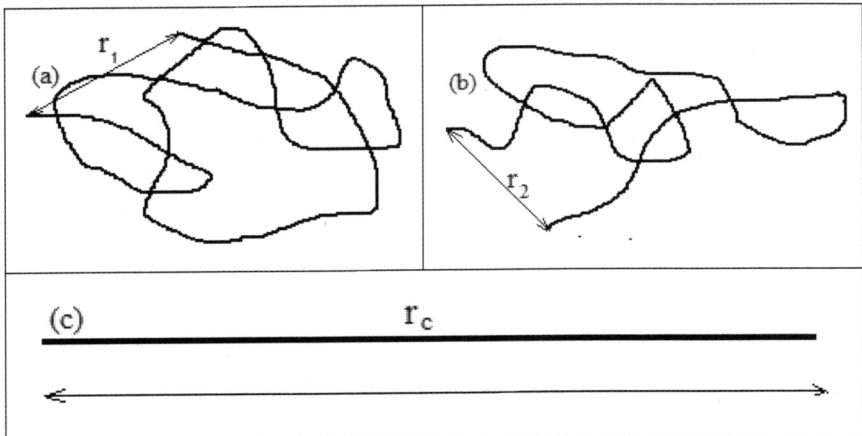

Figure 2.13   Schematic diagram of a polymer chain.

The distance between the ends of the straightened molecule is called the contour length (Fig. 2.13c). The contour length is related to the DP of the polymer. Each conformation of the polymer differs in size. The size can be characterized by end-to-end distance and the radius of gyration. The average end-to-end distance is the straight-line distance between two ends of the molecule as shown in Fig. 2.13a and b. The molecule can attain any conformation from the perfectly coil to perfectly straight line. So, as an average property, the end-to-end distance can be calculated as a root mean square end-to-end distance. This distance is basically square root of the average of the squared end-to-end distance and shown in Eq. (2.1).

$$R^2 = \frac{1}{P} \sum_{n=1}^{p} r_n^2 = \overline{r^2}.$$

(2.1)

where R is the mean square end-to-end distance, p the instantaneous conformations, r_n the end-to-end length of nth molecule. Polymers are free rotating chains, where the bond length is fixed, the bond angle is fixed and the dihedral angle is free. So the projected length of the ith segment on the straight line is $a \cdot \cos \theta$ and the sum of these projections is the end-to-end distance $\langle r \rangle$.

$$\langle r \rangle = \Sigma a \cdot \cos \theta$$

(2.2)

Consider a single isolated polymer chain, whose carbon atoms are linked by single bond. The extreme condition is of the polymer molecular chain with no rotation. The chain will be stiff and straight. The length will be equal to the product of the bond length and number of bonds present, as shown in Eq. (2.3).

$$\langle r \rangle = n \cdot a$$

(2.3)

The other extreme is the polymer chain where the valency angles are not fixed. The units rotate freely about the bond, and the chain can assume any conformations and is perfectly flexible. The root mean square distance $\langle r^2 \rangle$ may represent the molecular dimension for this flexible polymer chain. The root mean square distance may be estimated by means of random flight calculation and it is shown in Eq. (2.4)

$$\langle r^2 \rangle = n \cdot a^2$$

(2.4)

This expression is applicable to a polymer chain growing randomly in three dimensions. The real polymer chain consisting of carbon–carbon atom has additional restrictions like (a) fixed bond angle and (b) bond rotation. So the rotation is restricted because of these limitations. Hence, the real polymer chains will assume a smaller number of conformations.

2.13 Mathematical derivation of end-to-end distance

The mathematical derivation of Eq. (2.4) for the end-to-end distance as well as the restrictions imposed by bond angle and bond rotation can be computed mathematically. These are as follows:

(a) End-to-end distance with free rotation

End-to-end distance with free rotation can be computed by assuming the average size of the molecule expressed as mean square end-to-end length (R^2), as mentioned in Eq. (2.1) and Fig. 2.13.

$$R^2 = \frac{1}{P}\sum_{n=1}^{P} r_n^2 = \overline{r^2} \tag{2.1}$$

For a single chain, composed of N rigid bond connected ends to end and let the distance from the end to the end of the bond is vector $\vec{a}_j$, then

$$\overline{r}_j = \sum_{i=1}^{N} \overline{a}_i \tag{2.5}$$

Substituting Eq. (2.5) in Eq. (2.1) can modify shown in Eq. (2.6),

$$R^2 = \frac{1}{P}\sum_{n=1}^{P}\left(\sum_{i=1}^{N}\overline{a}_i\right)^2 \tag{2.6}$$

$$= \frac{1}{P}\sum_{n}^{P}(a_1^2 + a_2^2 + a_3^2 + a_4^2 + a_5^2 + a_6^2 + \cdots + a_N^2)$$

$$+ 2\vec{a}_1 \cdot \vec{a}_2 + 2\vec{a}_1 \cdot \vec{a}_3 + 2\vec{a}_1 \cdot \vec{a}_4 + \cdots + 2\vec{a}_1 \cdot \vec{a}_N$$

$$+ 2\vec{a}_2 \cdot \vec{a}_3 + 2\vec{a}_2 \cdot \vec{a}_4 + \cdots + 2\vec{a}_2 \cdot \vec{a}_N \tag{2.7}$$

$$+ 2\vec{a}_3 \cdot \vec{a}_4 + \cdots + 2\vec{a}_3 \cdot \vec{a}_N$$

$$+ \cdots$$

$$+ 2\vec{a}_{N-1} \cdot \vec{a}_N$$

For the square terms, all values are equal

$$a_1^2 = a_2^2 = a_3^2 = a_4^2 = a_5^2 = a_6^2 = \cdots = a_N^2 \tag{2.8}$$

Hence

$$R^2 = \frac{1}{P}\sum_{n}^{P}(Na_1^2 + 2\vec{a}_1 \cdot \vec{a}_2 + 2\vec{a}_1 \cdot \vec{a}_3 + \cdots + 2\vec{a}_{N-1} \cdot \vec{a}_N) \tag{2.9}$$

The quantity Na^2 is same for all molecules. Hence, the equation becomes,

$$R^2 = Na_1^2 + \frac{1}{P}\sum_n^p (\vec{a}_1 \cdot \vec{a}_2 + \vec{a}_1 \cdot \vec{a}_3 + \cdots + \vec{a}_{N-1} \cdot \vec{a}_N) \qquad (2.10)$$

The scalar product of $\vec{a}_1 \cdot \vec{a}_2$ equals to $a^2 \cdot \cos\theta$, where θ is the angle between the positive directions of the vectors $\vec{a}_1 \cdot \vec{a}_2$. For a random chain, there is no correlation between the directions of any links. So all values of θ are equally likely or $\cos\theta$ is equally likely be positive or negative. The sum over a large number of such product must be zero. So the average value of $\vec{a}_1 \cdot \vec{a}_2$ or any other similar pair will be zero. This means that the equation for freely orienting chains reduced to

$$R^2 = Na^2 \qquad (2.11)$$

Or the rms length

$$\langle r \rangle = N^{\frac{1}{2}} \cdot a \qquad (2.12)$$

Putting the value of $a = L/N$, where L is the total length, then

$$r = L / N^{\frac{1}{2}} \qquad (2.13)$$

Formally an effective length of link a_e can be defined as the value which leads to the correct prediction of the root mean square distance between chain ends

$$\overline{r^2} = N_e \cdot a_e^2 \qquad (2.14)$$

where N_e is the effective number of links of length a_e. In this equation, the restriction is that extended chain length $L = N_e \cdot a_e$. So we find,

$$a_e = \overline{r^2} / L \qquad (2.15)$$

$$N_e = L / \overline{r^2} \qquad (2.16)$$

rms length $= \langle r \rangle = N_e^{\frac{1}{2}} \cdot a_e \qquad (2.17)$

(b) End-to-end distance with bond angle restriction

The real chains are not freely orienting. The bond angles are fixed. A freely joining bond will create a sphere with respect to the first bond as it has no restriction of bond angle. On the other hand, the real chain will create cone with respect to the first bond. The schematic diagram of this difference is shown in Fig. 2.14. The end-to-end distance with restricted bond rotation can be computed by means of Eq. (2.10). In this equation, more detailed calculations can be made by presuming non-zero second term.

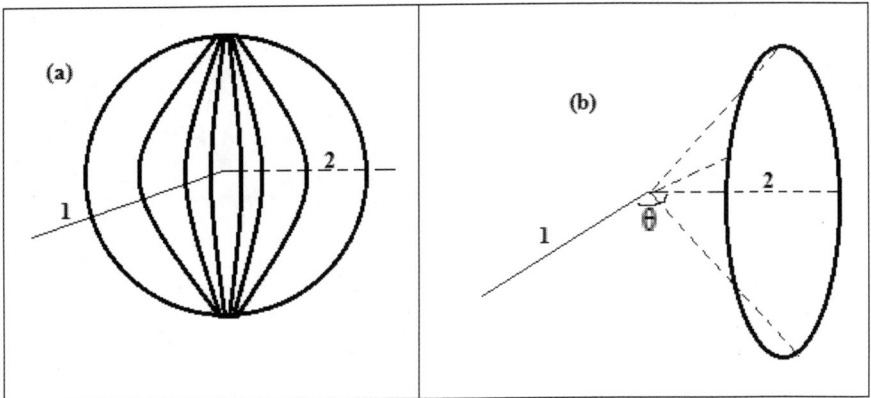

Figure 2.14 Schematic diagram of (a) freely joining ends and (b) ends joined with bond angle θ

For a chain of identical links, with no appreciable end effects, all terms as $\vec{a}_1, \vec{a}_2, \vec{a}_3, \vec{a}_4, \vec{a}_5 \cdots \vec{a}_N$ must have equal average values. In general any set of terms of the form $\vec{a}_i \cdot \vec{a}_{i+m}$ will have equal values. In a chain of N links, there are $(N - 1)$ ways of obtaining pairs of form $\vec{a}_i \cdot \vec{a}_{i+1}$ or in general, $(N - m)$ ways of obtaining pairs like $\vec{a}_i \cdot \vec{a}_{i+m}$. Adding up all the equations, will result in equation

$$R^2 = Na^2 + 2[(N-1)\vec{a}_i \cdot \vec{a}_{i+1} + (N-2)\vec{a}_i \cdot \vec{a}_{i+2} + \cdots + \vec{a}_i \cdot \vec{a}_{i+N-1}]_n \quad (2.18)$$

Considering a simple chain with free rotation about a fixed bond angle θ, each successive bond can take up a position around a cone of semi-angle α equal to $\pi - \theta$. So all adjacent links make an angle α with one another, and so the terms $\vec{a}_i \cdot \vec{a}_{i+1} = a^2 \cdot \cos\alpha$, $\vec{a}_i \cdot \vec{a}_{i+2} = a^2 \cdot \cos^2\alpha$ or in general $\vec{a}_i \cdot \vec{a}_{i+N-1} = a^2 \cdot \cos^m\alpha$. So Eq. (2.18) can be modified and rewritten as Eq. (2.19)

$$R^2 = Na^2 + 2[(N-1)a^2 \cdot \cos\alpha + (N-2)a^2 \cdot \cos^2\alpha \\ + \cdots + a^2 \cdot \cos^{N-1}\alpha]_n \quad (2.19)$$

$$= Na^2 + 2Na^2 \cdot \cos\alpha[1 + \cos\alpha + \cos^2\alpha \\ + \cdots + \cos^{N-2}\alpha]_n - 2a^2 \cdot \cos\alpha$$

$$[1 + 2\cos\alpha + 3\cos^2\alpha + \cdots + n\cos^{N-2}\alpha]_n \quad (2.20)$$

$$= Na^2 + 2Na^2 \cdot \cos\alpha[(1 - \cos^{N-2}\alpha)/(1 - \cos\alpha)] - 2a^2 \cdot \cos\alpha$$

$$[(1 - \cos^{N-1} \alpha) / (1 - \cos \alpha)^2] \tag{2.21}$$

In any polymeric material, it is known that N is the number of bonds and it is too large and $\cos \theta$ is not too close to 1.0. Hence, it can be assumed that (1) $\cos^{N-2} \alpha \equiv \cos^N \alpha \equiv 0$ (2) $1 - \cos^N \alpha \equiv 1$ (3) $N = N + 1$. So the equation will be

$$R^2 = Na^2 + 2Na^2 \cdot [(\cos \alpha) / (1 - \cos \alpha)] \tag{2.22}$$

$$= Na^2 [(1 + \cos \alpha) / (1 - \cos \alpha)] \tag{2.23}$$

or $$r = N^{1/2} \cdot a \cdot [(1 + \cos \alpha) / (1 - \cos \alpha)]^{1/2} \tag{2.24}$$

This equation can be modified with respect to the bond angle θ, and $\alpha = \pi - \theta$

or $$r = N^{1/2} \cdot a \cdot [(1 - \cos \theta) / (1 + \cos \theta)]^{1/2} \tag{2.25}$$

The value of $(1 - \cos \theta)/(1 + \cos \theta)$ is greater than unity as $\cos \theta$ value is negative, owing to the fact that bond angle is always greater than 90°. The value of r can be different when θ value is different as shown in following examples.

Example 1:   For bond angle $\theta = 90°$, then $\cos \theta = 0$ and $R = \sqrt{N} \cdot a$ (2.26)

Example 2:   For bond angle $\theta = 109\frac{1}{2}°$, then $\cos \theta = 0.34$ and
$$R^2 = N \cdot a^2 \cdot (1.34/0.66) \text{ or } R = \sqrt{(2N)} \cdot a \tag{2.27}$$

The average of the cosines of the orientation angle, $\langle \cos \theta \rangle$ is equal to the ratio of the actual end-to-end distance to the fully extended end-to-end distance.

$$\langle \cos \theta \rangle = r / n \cdot a = [\Sigma a \cdot \cos \theta] / n \cdot a = \Sigma \cos \theta / n \tag{2.28}$$

This equation means that for a given and specified polymer, $\langle \cos \theta \rangle$ changes only if the end-to-end distance changes.

(c) End-to-end distance with bond angle and bond rotation angle restriction

Another important factor is the bond rotation. As seen from Fig. 2.14, the bond angle restricts the movement of the second bond with respect to the first bond with an angle θ with the first bond. This will form a cone if all the probable points of the bond will be recorded. A bond rotation angle will further restrict the movement of the bonds. The schematic diagram of this bond rotation is shown in Fig. 2.15. In this case, the movement of the third bond with respect to the first bond is restricted because of mutual attraction and/or repulsion.

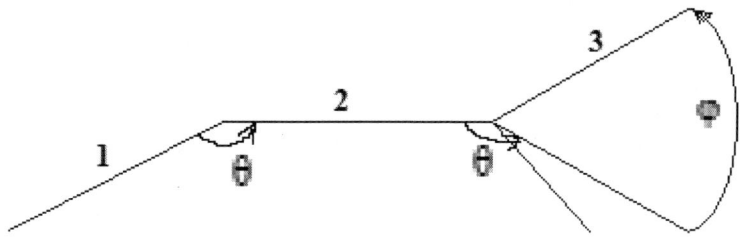

Figure 2.15   Schematic diagram of polymer molecular chain with bond
angle θ, and bond rotation angle ϕ.

In this case, minimum energy calculation will be applied for the conformation of the molecular chain, If the bond rotation angle will be put as ϕ, then the equation can be further modified as

or $$r = N^{1/2} \cdot a \cdot [(1-\cos\theta)/(1+\cos\theta)]^{1/2} \cdot [(1+\cos\varphi)/(1-\cos\varphi)]^{1/2} \quad (2.29)$$

For example, with configuration out of infinite positions, or 360° positions, at every ϕ angle, with bond angle θ, there will be only three (3) minimum energy position. These are trans, ($\phi = 0$) gauche ($\phi = 120°$) and gauche ($\phi = 240°$) conformations. The value of $[(1+\cos\phi)/(1-\cos\phi)]^{1/2}$ is always greater than unity.

2.14   Radius of gyration

This is the parameter which can be defined as the size of the molecule. This is generally noted as 'S'. Qualitatively, this is the average distance of the mass in a molecule from the centre of the mass. More precisely, it is

$$S^2 = (1/N)\sum_{i=1}^{N} S_i^2 \quad (2.30)$$

where S_i is the distance of the ith element from the centre of mass

For random chain,

$$S^2 = l^2/6 \quad (2.31)$$

or $$S = r/\sqrt{6} = \sqrt{N}/\sqrt{6} \cdot a \quad (2.32)$$

2.15   Excluded volume effect

The actual conformations taken up by polymer chains in solution will depend upon the following factors:

1.   The strength of interaction between the molecules, i.e. strong interaction between polymer chains and weak interaction between solvent polymers will result a fairly tight conformation and occupy less volume than a conformation calculated.

2.   The chains have changes or mutually repulsive groups and the chains will tend to avoid curled up position, a more open conformation.

3.   Conditions of the solution, i.e. temperature, pressure, etc.

If all the factors are balanced, then the polymer chains would take up theoretically predicted positions

Further molecules take up space. Many of the theoretical conformations would not be possible because they would demand simultaneous occupation of the same space by different part of the chain. This is known as 'Excluded volume effect', i.e. the real polymer conformation should be more open than those predicted theoretically.

2.16   Role of solvent on conformation

The arrangement of the chain segments within small domains or regions of space and configurational characteristics of the polymer chains expressed in their mean dimensions are inseparably related. The dimensions of an isolated, unperturbed chain as a whole can be determined in a theta solvent for a perfectly flexible. The solvents depending upon its degree of interaction with the particular polymer can penetrate inside the structure and loosen the structure. Because of the loosening of the structure, the conformation of the chain molecules changes. Brief summary of the action of different types of solvents on chain conformations is shown in Table 2.5.

A poor solvent will not change the conformation and the conformation will be tight. On the other hand, a good solvent will change the conformation drastically and convert the conformation to an open conformation. An ideal conformation can be possible by means of theta solvent. Alternately theta solvent is a solvent for a particular polymer which gives an ideal conformation, where the measured end-to-end distance is equivalent to the theoretical value. In general, for all chemical activities, theta solvent is referred.

Table 2.5   Effect of solvent on chain conformation

Type of solvent	Interaction between		Effect on polymer conformation
	Solvent-polymer	Polymer-polymer	
Poor	Weaker	Stronger	Tight conformation
Moderate	Equal	Equal	Close to ideal conformation
Theta	Slightly stronger	Slightly weaker	Ideal conformation, interaction with solvent just balances excluded volume
Good	Stronger	Weaker	Open conformation

2.17   Conformation and chain flexibility

For any constant value of a and n, the increase in molecular dimensions indicates that there is a decrease in chain flexibility. The molecular chain shows the tendency of forming more extended conformation. The chain flexibility is also affected by chemical structure of the polymer molecule. Some of the important parameters which influence chain flexibility are as follows:

1.   Non-symmetric substitution of carbon backbone chain,

2.   Presence of bulky substituent groups,

3.   Strongly interacting substituent groups like polar groups,

4.   Presence of any other atoms other than carbon atoms in the backbone,

5.   Presence of aromatic rings like benzene ring,

6.   Presence of cyclic structures in the main chain,

7.   Presence of cyclic structures in the side chain,

8.   Presence of intramolecular interactions, which give rise to helix type of structures.

2.18   Chain structure in selected fibres/polymers

2.18.1 Polyethylene

Polyethylene has a basic structure of $-CH_2-CH_2-$. This consists of a long chain of repeated ethylene units. Even though, polyethylene has a simple structure, but there are several types of polyethylene with different branching. Based on branching, polyethylene can be classified as high density polyethylene (HDPE), high molecular weight high density polyethylene (HMHDPE), low density polyethylene (LDPE), linear low density polyethylene (LLDPE) and very low density polyethylene (VLDPE). HDPE is related to that linear polymeric chain having no or very few short side chains. This polymer is higher crystalline (70–80%) with a density of up to 0.96–0.97 g/cm^3. Higher molecular weight polyethylene with no side chains is denoted as HMHDPE. This type of polyethylene is used for high performance polyethylene fibres. To reduce crystallinity, other comonomers are introduced with short side chains. This polymer is called LLDPE and has a lower crystalline (40–50%) with a density of up to 0.94–0.96 g/cm^3. Depending upon the comonomers, and formation of side chains the polymer can be LDPE or VLDPE, where the density is in between 0.915 and 0.94 g/cm^3 or 0.88 and 0.912 g/cm^3, respectively. The schematic diagram of this polyethylene is shown in Fig. 2.7.

2.18.2 Proteins

Proteins are polymeric substances with relative molecular masses of many thousands and made of amino acids. Wool, silk, casein are different proteins used as fibres. There are about 20 different amino acids in proteins. Each fibre consists of different amino acids in different proportions. The majority of the amino acids have the general formula NH_2–CHR–COOH or $^+NH_3$–CHR–CO^-. When the amino group of one molecule condenses with the carboxylic group of the second molecule, a dipeptide is formed, like

NH_2–CHR_1–CO–NH–CHR_2–COOH   or $^+NH_3$–CHR_1–CO–NH–CHR_2–CO^-·A

The peptide bond is formed via covalent binding of the Carbon atom of the carboxyl group of one amino acid to the nitrogen atom of the amino group of another amino acid. Condensation with a further amino acid gives a tripeptide and the process continues to form a polypeptide.

So the polypeptide molecules consist of different side groups. Because of the spatial distribution of the groups around carbon atom, R groups are placed in different manner at different places. The distribution and nature of this side group is an important factor for the structure and properties of each individual protein. Also, each protein molecule has different types of groups or side groups (R). Small size of R will arrange the molecule to be more linear, whereas presence of bigger groups in side will induce the molecule in a similar manner like that of a branched chain. So the molecules in protein can be arranged either in (a) straight chain, or (b) folded chain or (c) coiled chain structure. The schematic diagram of these structures is shown in Fig. 2.16. Further, different types of forces like hydrogen bonding, salt linkages and cystine linkages make the molecules firm and rigid. The forces are acting either as intermolecular forces or intramolecular forces.

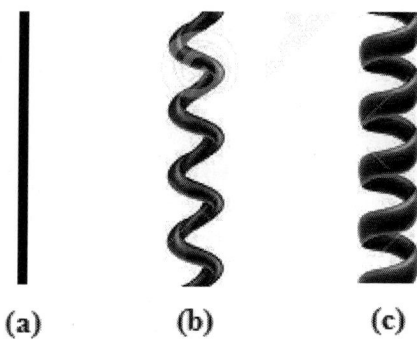

(a)   (b)   (c)

Figure 2.16   Structure of protein molecules (a) straight chain, (b) folded
                chain (c) coiled chain.

(a)   Straight chain

For protein molecules, simple and short side chains will permit the formation
of a fully extended chain. For example, polyglycine or the polypeptide
molecules having higher amount of glycine will form a fully zigzag extended
chain. Silk fibroin forms an extended zigzag chain molecule due to the
presence of higher amount of glycine. The protein molecule in this extended
form has the property of adhering to adjacent chains.

(b)   Helical chain

For fibrous protein, like keratin, the side chains are more complex, which
restricts to form a fully extended polypeptide chain molecule. Owing to this
complexity, the protein forms a folded chain to accommodate the side chains.
In case of folded chains, it is possible to unfold these chains and prepare a
straight chain structure. The normal form of wool fibre is based on folded
molecular chains. This form is generally referred to as   keratin. The structure
can be stretched to twice its original length to another structure, known as
keratin. Also, the structure can be reversed back on contraction. In case of
keratin, the more complex side chains restrict its full extension.

(c)   Coiled chain

The polypeptide chains of globular proteins are more tightly coiled. The
structure may be pictured to be one in which the long polypeptide chains
are folded in definite planes, in which themselves are arranged in a coiled
structure. There are 3.5 amino acid residues in each turn of the coil. The coil
spacing is 0.54 nm. The length of each amino acid residue in projection on
the longitudinal axis is 0.156 nm. There is no interaction between the chains
and so there is no intermolecular force between two coiled chain molecules.
As there is no interchain entanglement, the polypeptide molecule exists as
relatively small aggregates. These types of molecules can be completely

surrounded by solvent molecules. So globular proteins are readily soluble in water or dilute aqueous solutions of inorganic salts. The coiled chains can also be uncoiled. The fibrous and globular forms of protein are interchangeable. The transformation of molecules of coiled or folded chains in solution to the insoluble and extended form is referred to as 'denaturation'.

Further readings

1. Shaw, M. T., *Introduction to Polymer Viscoelasticity*, Wiley Interscience, 1985.

2. J. J. Aklonis, W. J. Mac-Knight, and M. Shen (Eds.), *.Introduction to Polymer Viscoelasticity*, John Wiley & Sons, Inc., New York, 1972.

3. Robert J. Young and Peter A. Lovell, *Introduction to Polymers*, CRC Press, 2011.

4. J. W. Hearle and R. H. Peters (Ed.), *Fibre Structure*, Butterworth, London, 1963, p. 346.

5. W. E. Morton and J. W. S. Hearle, *Physical Properties of Textile Fibres*.

Crystalline structure

3.1 Introduction

Polymer molecular chains are linear, flexible and so can be arranged in a close packing. The close packing lead to a highly ordered microscopic structure and referred as 'crystalline'. In crystallography, the crystalline structure or crystal structure is a unique arrangement of atoms, ions, molecules and so it describes a highly ordered structure due to its intrinsic nature. Crystalline structure is that structure which exhibits a typical x-ray diffraction pattern.

The most stable arrangement or ordered arrangement can be achieved by proper placement of atoms or atomic groups in like manner. Because of the valence angle of the carbon atoms, the molecular chains are placed either in a zigzag manner or in a helical manner. For an assemblage of close rod like molecules of zigzag structure, the closest packing can be achieved by stacking the rods with their axes parallel to each other.

3.2 Intermolecular forces

Many polymer molecules contain atoms or groups, which interact with other, where similar or dissimilar atoms or groups present in neighbouring molecular chains. This leads to formation of intermolecular forces. It may be Coulombic forces, ion–ion interactions, ion–dipole interactions, dipole–dipole interactions or connected with the formation of H-bonds. The simplest is van der Walls forces. **Intermolecular forces** are forces of attraction or repulsion which act between neighbouring particles (atoms, molecules or ions). They are weak compared to the intramolecular forces, the forces which keep a molecule together. For example the covalent bond, involving the sharing of electron pairs between atoms is much stronger than the forces present between the neighbouring molecules.

For these types of interactions, the interactive groups must be within a specified distance from each other. To obtain sufficient proximity between strongly attractive groups, each group should extend from the molecular chain axis in the direction of the neighbouring chain. Strong interchain attractions as well as strong interchain repulsions of the interactive groups can be satisfied if only these interactive groups are suitably oriented and placed in a parallel manner with respect to the chain axis.

Intermolecular forces have four major contributions:

1. A repulsive component prevents the collapse of molecules.

2. Attractive or repulsive electrostatic interactions between permanent charges, dipoles, quadruples and in general between permanent multipoles.

3. Induction forces or polarization forces, which is the attractive interaction between a permanent multipole on one molecule with an induced multipole on another.

4. Dispersion or London dispersion forces, which is the attractive interaction between any pair of molecules, including non-polar atoms, arising from the interactions of instantaneous multipoles.

The intermolecular forces between the molecular chains are exclusively of van der Waals forces. The van der Waals force is caused by temporary attraction between electron rich regions of one molecule and electron poor regions of another. These attractions are very common but are much weaker than chemical bonds. These forces are relatively weak and act between all pairs of atoms and tend to produce a close packed structure. The van der Waals force is the sum of the attractive or repulsive forces between molecules (or between parts of the same molecule) other than those due to covalent bonds, or the electrostatic interaction of ions with one another, with neutral molecules or with charged molecules. The term includes:

*   Force between two permanent dipoles,

*   Force between a permanent dipole and a corresponding induced dipole,

*   Force between two instantaneously induced dipoles (London dispersion force),

*   It is also sometimes used loosely as a synonym for the totality of intermolecular forces. van der Waals forces are relatively weak compared to covalent bonds.

Van der Waals forces are usually two types like polarization forces and dispersion forces. The polarization forces are the attraction of molecules by permanent dipoles and by induced dipoles. The hydrophobic interaction is the interaction of hydrophobic groups, especially alkyl chains to associate together and escape from an aqueous environment.

A hydrogen bond is the electrostatic attraction between polar molecules that occurs when a hydrogen (H) atom bound to a highly electronegative atom such as nitrogen (N), oxygen (O) or fluorine (F) experiences attraction to some other nearby highly electronegative atom. This bond is a strong electrostatic dipole–dipole interaction. Like covalent bonding, it is directional, stronger than a van der Waals interaction, produces interatomic distances shorter than the sum of van der Waals radius and usually involves a limited number of interaction partners, which can be interpreted as a kind of valence. These hydrogen-bond attractions can occur between molecules (*intermolecular*) or within different parts of a single molecule (*intramolecular*). The hydrogen bond (5–30 kJ/mole) is stronger than a van der Waals interaction, but weaker than covalent or ionic bonds.

So in a polymer molecule, hydrogen bonds occur mainly with hydrogen atom bound to nitrogen, oxygen or carbon. The required condition is that the hydrogen atom must be covalently bonded to one of high electron affinity, whereby it is positively polarized. It can act five different ways: (i) intermolecularly and extending over many molecules, (ii) intermolecularly so that two molecules are united in a dimer, (iii) intermolecularly where the electron donation is by a double bond by ϖ-electron system, (iv) intramolecularly with ring formation and (v) intramolecularly in an ion. These forces are directional. In some polymers, there will be chemical cross-links like disulphide bonds or ester linkages. These intermolecular interactions are interchain cross-links. If interchain cross-links exist, it will impose serious restrictions on the parallelization of the cross-linking groups relative to the chain axis.

3.3 Crystal structure

The polymer or the fibre molecular chains consist of identical and reasonable symmetrical units. The molecular chains or parts of it may arrange themselves in a regular or crystalline arrangement, extending over considerable distances in the direction of the chain axis. This also develops intermolecular forces between the active sites within specified distances between chain molecules and restricts the bond rotation. This develops more ordered arrangement in the structure and termed as 'crystalline state' and the materials as 'crystallite'. Crystalline structure deals with the organization of large chain molecules in highly ordered three-dimensional structures.

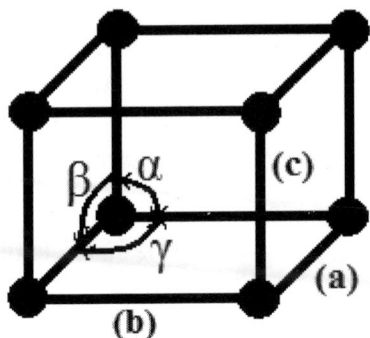

Figure 3.1   Unit cell parameters.

The smallest repeat unit of the three dimensional lattice of the crystallites is known as 'crystal' or 'unit cell'. Unit cells are made by defining a set of coordinate axes composed of three non-coplanar, non-collinear lattice vectors. A primitive unit cell contains one lattice point, but may contain many atoms. The crystals are characterized by three axial dimensions, i.e. *a, b* and *c* and three angles, i.e. α, β and γ. A schematic diagram indicating the parameters is shown in Fig. 3.1. Each crystal system has its own parameters as mentioned in Table 3.1. The crystals are divided into seven systems according to the form of a parallelepiped with their own constants.

Table 3.1   Crystal parameters

	Crystal types	**Sides**	**Angles**	**Parameters**
1	Cubic	$a = b = c$	$\alpha = \beta = \gamma = 90°$	a
2	Tetragonal	$a = b \neq c$	$\alpha = \beta = \gamma = 90°$	$a\ c$
3	Hexagonal	$a = b \neq c$	$\alpha = \beta = 90°, \gamma = 120°$	$a\ c$
4	Orthorhombic	$a \neq b \neq c$	$\alpha = \beta = \gamma = 90°$	$a\ b\ c$
5	Trigonal	$a = b = c$	$\alpha = \beta = \gamma \neq 90°$	$a\ \alpha$
6	Monoclinic	$a \neq b \neq c$	$\alpha = \gamma = 90° \neq \beta$	$a\ b\ c\ \beta$
7	Triclinic	$a \neq b \neq c$	$\alpha \neq \beta \neq \gamma \neq 90°$	$a\ b\ c\ \alpha\beta\gamma$

The most simple crystal system is a cubic system, having only one parameter, i.e. one side or '*a*'. In this system, each side and each angle are equal and the angles are equivalent to 90°. The tetragonal is a square prism and so only two sides are the parameters like '*a*' and '*c*'. The hexagonal is a prism on a 60° parallelogram and so denotes by two parameters like tetragonal. The orthorhombic is a rectangular parallelepiped and denoted by all three sides. The trigonal is a cube deformed along one diagonal and so only one side, '*a*' and one angle is required as its parameters. The monoclinic system is a right prism with parallelogram as base and so it is notified by all three sides and one angle. The extreme crystal system is the triclinic system, where each side and each angle are unequal. It is a general parallelepiped. The crystal structures of some of the fibres are discussed below.

3.3.1 Crystal structure of cellulose

Cellulose

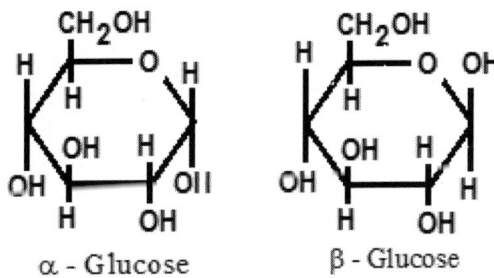

α - Glucose   β - Glucose

Figure 3.2   Structure of α-Glucose and β-Glucose.

The basic unit of cellulose is glucose. The structure of the glucose unit is shown in Fig. 3.2. This is a pyranose ring. There are two isomers of glucose and it differs in the arrangement of the CHOH group. The structure of the cellulose is shown in Fig. 3.3. The repeat unit of cellulose is the cellobiose. Cellobiose is β-glucosidogluco pyranose, formed by two units of glucose. β (1–4) linked anhydro glucose units, with a rotation of 180° around the chain axis for subsequent ring unit forms the cellobiose unit. The identity period has to be at least two or a multiple of two glucose units long. The molecule itself is stiff and straight because of its internal hydrogen bonds and the α linkage. The non-bonded hydroxyl groups on a single molecule will tend to form hydrogen bonds with acceptor compounds coming nearer than 3 Å. If the ring planes are parallel to each other, the elementary fibril is constrained by three different forces, which tend to hold the cellulose structures together. These forces are:

(a)   The main valence bonds, i.e. covalent bonds in the main chain,

(b)   Hydrogen bonds extending sideways from the rings and

(c)   Dispersion forces perpendicular to ring planes.

Figure 3.3   Structure of cellulose molecule.

Table 3.2 Crystal structure of cellulose

	Cell I	Cell II	Cell III	Cell IV
Crystal system	Monoclinic	Monoclinic	Monoclinic	Monoclinic
a(Å)	8.35	8.14	7.74	8.12/7.90
b(Å)	10.3	10.3	10.3	10.3
c(Å)	7.90	9.14	9.9	7.99
β (°)	84/83.3	62	58	90
Density (g/cm3)	1.625	1.583	1.620	1.610

Cellulose molecules have the ability to crystallize into different polymorphic forms. These polymorphic forms are noted as cellulose I, cellulose II, cellulose III, cellulose IV, as shown in Table 3.2. The elementary crystal unit cell is monoclinic and belongs to the space group $P2/1$, in which two sets of diagonal screw axis are present. The type of cellulose exhibiting this structure is termed as cellulose I. Cellulose I exists in native cellulose like cotton, flax, jute, hemp, sisal, coir, etc. When swollen in strong swelling agents, a shifting of the a–b plane occurs, leading to formation of cellulose II. Cellulose II is the common form found in rayon, cellophane. Mercerization of native cellulose I leads to formation of cellulose II. The weakest cohesion forces, i.e. dispersion forces between the hydrocarbon conformations, determine the first structural order of cellulose in nascent state as well as during the precipitation of regenerated celluloses, since the stronger hydrogen bonding forces are shielded by water layers. Certain treatments will transform cellulose I and cellulose II into cellulose III and cellulose IV. Cellulose III is a low temperature modification of cellulose I and it is formed by treatment of cellulose I or cellulose II in liquid ammonia or in dry ethylamine. Cellulose IV is a high temperature phase modification of cellulose I or cellulose II at high temperature. This form of cellulose along with cellulose II is present in those regenerated cellulosic fibres, where the fibres formation undergoes high temperature treatment like high weight modulus fibres, or tyre cords fibres. Simply heat treatment cellulose also forms cellulose IV. Cellulose IV can be formed on heating cellulose II in glycerol at 200°C. Cellulose I will be converted through cellulose III, whereas cellulose II to directly cellulose IV. Native celluloses have microfibrillar widths of 200–300 Å and the same for regenerated celluloses is around 50–100 Å. Cotton structure is one phase crystalline structure with accessible regions of limited disorder. The extent of hydrogen bonding in the accessible region varied from near perfection to almost complete disorder. Perfection of hydrogen bonding decreases with decrease in crystallinity. The regenerated cellulose is characterized by cellulose II structure with folded structure. A single chain of cellulose is folded back and forth to form a sheet like structure.

3.3.2 Crystal structure of silk

Silk fibre consists of two types of protein macromolecules in its structure. The major part is fibroin and it is surrounded by glue, known as sericin. The fibroin is a semicrystalline structure. The crystalline region occupies around 20–25% of silk. The macromolecule in fibroin in the solid filament is α-polypeptide structure and it is based on extended chain molecule. The side chains are simple, which permit fully extended main chain. The schematic structure of fibroin is shown in Fig. 3.4.

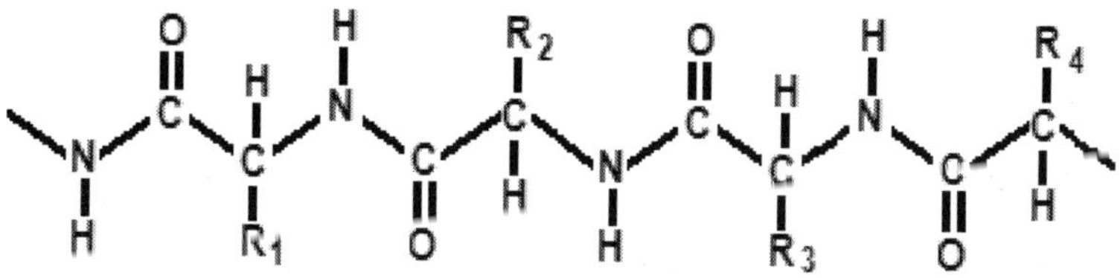

Figure 3.4   Structure of silk fibroin.

The amino acid composition of the crystalline fraction of silk fibroin is different from that of amorphous fraction of silk fibroin. The acid composition of crystalline fibroin is Glycine (GLY), 0.48; Alanine (ALs), 0.33; Serine (Ser), 0.15; and Tyrosine (Tyr), 0.01; by fraction and the acids (Ala, Gly, Ser) are in the ratio 3:2:1. The molecular structure of the silk fibroin, with three amino acids Ala, Gly and Ser, has a 1/1-helical symmetry. This means that two repeating units of Ala-Gly are constrained in the fibre repeating period. The geometry for Ala is for the D-Ala and not L-Ala. In the crystalline fibroin, the sequence of glycine occurs alternatively with alanine or serine and stretched to full extent (R_1, R_3 positions in Fig. 3.4). The other amino acids, like tyrosine, lysine and arginine because of their larger side groups are present in amorphous region only. The unit cell of the crystalline structure of silk crystalline fibroin is an orthogonal cell. The cell parameters of this structure is $a = 9.40$ Å, $b = 9.20$ Å, c (fibre axis) $= 6.97$ Å with $b = 90°$. The space group is $P21$. In order to pack four chemical repeating units in a unit cell the helical axis of the molecule must coincide with the c-axis which is parallel to 21-axis. The unit cell contains four molecular chains. Structurally the unit cell contains two alanine units and two glycine units. The space group and the number of a repeating unit of (Ala-Gly)$_2$-Ser-Gly in a unit cell were determined to $P21$ and four, respectively. Higher amount of glycine and alanine imparts stability to the chain and creates possibility of high orientation. In this extended form of the main chain, CO-NH groups of the successive residues protrude to opposite sides and form intermolecular hydrogen bonds. It has a two-fold screw axis with a repeat distance of around 7.32 Å. All the side chains of the different amino acids are all extended from the main chain. The chains are arranged in an anti-parallel and orderly manner.

In general, there are four models for the sheet structure formed by hydrogen bonds. They are: (i) polar-anti-parallel [PA (1)]; (ii) polar-parallel (PP); (iii) anti-polar-anti-parallel (AA); and (iv) anti-polar-parallel (AP). In the polar model, the methyl groups of alanine residues are on one side of the sheet only, while in the anti-polar model, the methyl groups alternately point to both sides of the sheet along the hydrogen bonding direction. The crystal structure of silk usually corresponds to PA sheets. Hence the fibroin crystalline structure may be built up in which anti-parallel chains are linked together to form sheets. The hydrogen bond forms the glycine residues protruding from one side of the sheet and hydroxymethyl and methyl groups protrude from the other side of the sheet. Hydrogen bonds are generally formed between CO and NH groups of adjacent chains.

The intrachain weak hydrogen bonding network is formed in between the O and N atoms of glycine. The intermolecular spacing is around 2.73 Å and the bond angles of 161.52° whereas the intrachain hydrogen bonding N–H···O and N–H···O in alanine are at bond distances of 2.74 and 3.06 Å, respectively. The corresponding bond angles are 165.21° and 136.02°. Further the intramolecular distances between O atom of serine residue and O atom of alanine residue is 2.83 Å and the bond angle is 89.54°. This may suggest that the OH groups of the serine residues are associated with the hydrogen bonding network forming the sheet structure. The formation of hydrogen bond can be explained with the concept of strong, moderate and weak on the basis of observed hydrogen bonding geometries. The separation distances are 1.2–1.5 Å, 1.5–2.2 Å and 2.2–3.2 Å for strong, medium and weak bonds and the bond angles are 175–180°, 130–180° and 90–150°, respectively. Most of the hydrogen bonds are medium or weak. The chain conformation is stabilized by the bifurcated hydrogen bond between N of alanine and serine residues. Because of these hydrogen bonds, the molecular conformation is restricted in its degree of freedom by forming the β-pleated structure. The hydrogen bonds are the only direct interaction between adjacent layers other than van der Waals interaction. There is fairly strong intramolecular hydrogen bond among Ala, Gly and Ser residues.

The molecular conformation is essentially the pleated sheet structure. The molecular conformation must satisfy both sterical and mathematical requirements. The sheet structures formed by hydrogen bonds assume the anti-polar-anti-parallel arrangement. The arrangement of the molecular chains with hydrogen bonds can be described as a pleated sheet. van der Waals forces or intermolecular attraction are formed between individual polypeptide chains in the third direction. Individual chains are cross-linked by H-bonds and by van der Waals forces. The pleated sheets are packed regularly and form a regular three-dimensional order. The distribution of pleated sheets along c axis are separated by 3.5 Å and 5.7 Å. The side chains of the amino acid project perpendicularly and the side chains of adjacent residues protrude on opposite sides within a given chain, the packing distance of H-atoms is 3.5 Å and for methyl or hydroxymethyl group is 5.7 Å.

The structure of the relaxed and the strained fibre can be studied by means of the normal silk fibre and oriented silk fibre. The oriented samples of silk fibre can be prepared by maintaining the freshly extracted silk gland with dilute acetic acid and then immediately deforming the contents of the gland by stretching and rolling, which leads to doubly oriented samples for x-ray recordings. The amino acid sequence in silk in stretched condition is Gly-L-Ala-Gly-L-Ala-Gly-[L-Ser-Gly-(L-Ala-Gly)$_n$]$_8$-L-Ser-Gly-L-Ala-L-Ala-Gly-LTyr where $n = 2$. The difference between the relaxed and strained fibers lies in the fact there are conformational changes in the chain either from *gauche* to *trans* in certain portion of the chains. The structure of fibroin is similar to those of synthetic polypeptides. The crystal structure of mulberry silk and wild silk are similar having fractions of crystalline region. The repeat distance in the direction of the fibre axis is same and also in the formation of pleated sheet, i.e. formation of H-bonds is similar. There is a slight difference between the packing of the pleated sheets because of the differences in the projection of side chains. The dimension is 10.0–10.6 Å instead of 9.29 Å.

3.3.3 Crystal structure of wool

Wool fibre is a protein fibre consisting of polypeptide structure. The three-dimensional structure of proteins is uniquely determined by its primary structure. The protein structure showed the polypeptide chain bundled into a compact tertiary structure. There are two types of protein structures, i.e. α-helix structure and β-structure. There is a regular helical arrangement of the polypeptide chain molecules and it is known as α structure. The term helical means that the molecules repeat itself periodically in the axial direction. In this structure, the polypeptide chain coils in a right-handed manner, the CO and NH groups of residues ith and $(i + 4)$th, respectively, form hydrogen bonds and it stabilize the helix. All or most of residues in a helix are bonded in this way, making it a relatively rigid unit of structure with very little free space in its core. The helix makes a whole turn per 3.6 residues or 18 residues in 5 turns, over an axial length of 2.7 nm or 0.15 nm/residue. The total chain can have between 4 and around 50 residues.

The most regular form of extended polypeptide chain seen in protein structures is the β-structure. This structure is stabilized by means of intermolecular interactions with the neighbouring chain molecules. This leads to formation of a sheet like structure. Like helices, sheets too are stabilized by hydrogen bonds between CO and NH groups, but in this case they are between chain molecules. The chain molecules can be both parallel and anti-parallel.

The polypeptide chain can make very sharp changes in direction using as few as four residues. Due to the geometry of the peptide backbone, the amino acid side chains exist alternately on either side of the sheet.

Wool fibre consists of a special type of polypeptide chain, known as keratin with active sites for the formation of all types of interchain bonds like hydrogen bonds, hydrophobic bonds, van der Waals forces, salt linkages and cystine linkages because of different types of amino acids. The schematic structure of wool keratin is shown in Fig. 3.5 and it is present in two forms. One is a normal form and another is a stretched form. Normal wool fibre structure is based on folded molecular chains and termed as α-keratin, whereas the extended form as β-keratin. The normal unstretched wool fibre is capable of being stretched to twice its original length or of being contracted to half the original length. α-keratin is regarded as at intermediate form being between fully extended and fully contracted modification. Normal wool fibre structure imparts the properties of long-range elasticity.

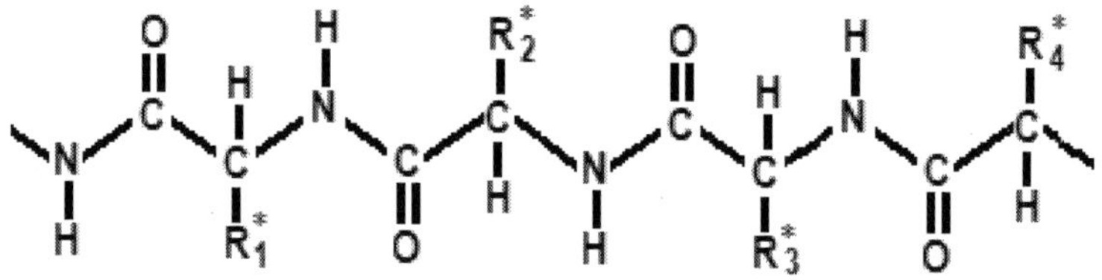

Figure 3.5   Structure of wool keratin with active sites (*).

With α-keratin, the molecular repeat period occurs at intervals of 5.1 Å and with β-keratin, it is 10.2 Å. The periodicity occurs every 3-amino acid residue. The repeat of each amino acid in extended form is 3.4 Å. The fold of the polypeptide chain consists of approximately 50% of the average length of the α-keratin unit, i.e. 1.7 Å. On the other hand, this repeat period in silk is 3.5 Å. The small differences are due to a slight fold in β-keratin as a result of more complex side chains. These side chains hinder the full extension of the main polypeptide chain. Further, there will be formation of cystine bridges and salt linkages at the active sites R_1, R_3 positions in Fig. 3.4, provided there will be the presence of exact type of amino acid groups. These interlinks between the chains impart lateral cohesion and rigidity to the looped formation of the main chain.

In β-keratin, there are two sets of planes parallel to the fibre axis and at right angles to one another. One plane gives a spacing of 4.5 Å and the other 9.8 Å. The distance of 4.5 Å is the spacing of the peptide chains from each other. In this plane, the peptide chains are separated by hydrogen and oxygen atoms with a space of around 1.65 Å between the centres of hydrogen and ketonic oxygen. This permits hydrogen bonding. The distance of 9.8 Å is the spacing of the planes in which the peptide chains lie. It is determined by the length of the side chains, which are situated at right angles to the plane of the peptide chains. This spacing is common to both α- and βkeratin. The cross-linkages are sufficiently firm to permit folding and extension and so play at important part in the elastic properties of the fibres. Because of these complications, the wool keratin is considered to exist in the form of a grid structure.

3.3.4 Crystal structure of polyethylene and polypropylene

Polyethylene and polypropylene fibres are carbon-chain polymers because of presence of only carbons in their main chain. Because of carbon atom, the chain molecules are flexible and can be arranged in any arrangement including folded chain or extended chain. Polyethylene fibre structure consists of only one type of structure and it is orthorhombic type. The unit cell of the crystalline structure of polyethylene fibre structure is $a = 7.42$ Å, $b = 4.94$ Å and $c = 2.55$ Å. The linear (unbranched) polyethylene was 80% or 90% crystalline. The crystals can be formed by precipitation from hot solutions in xylene. These crystals were in the form of *lamellae*, namely flat sheets several microns in lateral extent (a and b axes) but very thin (10 nm along the c-axis). The crystalline structure of polyethylene shows chain folding, with several microns long chain molecules, repeatedly folding back and forth through the crystals in a conformation. Thin lamellae forming complicated dendrites are formed from solutions at low temperature. On the other hand, simple thin lamellae are formed from solutions at high temperature. Bulk crystallization leads to medium thickness lamellae or thick lamellae. At low temperature of crystallization, medium thickness lamellae with an S-profile can develop into banded spherulites and on the other hand high temperature crystallization forms thick flat lamellae, and slowly branching to form spherulites.

Isotactic polypropylene, iPP has a particularly complicated crystalline microstructure, which depends, amongst other factors, on the mechanism and rate of crystallization. Four different crystalline structures have been described, corresponding to monoclinic (α), hexagonal (β), triclinic (γ) and smectic or quenched polymorphs. The α structure is a monoclinic structure with the dimensions $a = 6.65$ Å, $b = 20.96$ Å, $c = 6.50$ Å and $\alpha = 99.2°$. The crystallographic density of this structure is 0.936 g/cm³. β structure is a hexagonal structure and the γ structure is a trigonal structure. The α monoclinic structure being that principally obtained under typical industrial and laboratory processing conditions. The β-form occurs more rarely than the α-form since it is thermodynamically less stable, although under given thermal conditions it can occur simultaneously, and its appearance can be favoured under conditions of shear stress. The γ trigonal structure consist of the following dimensions $a = 6.54$ Å, $b = 21.4$ Å, $c = 6.5$ Å, $\alpha = 89°$, $\beta = 99.6°$, and $\gamma = 99°$. The crystallographic density of this structure is 0.954 g/cm^3. Rapid quenching of the fibre produces a smectic melt, but not completely amorphous. Cooling of the fibre under different conditions produces complex structures composed of α, γ and smectic structures.

3.3.5 Crystal structure of Polyamide 6

Polyamide 6 is polymorphic in character with different crystalline modifications. The different structural forms of Polyamide 6 consist of amorphous 1, amorphous 2, γ-pseudo hexagonal, β-hexagonal, α-paracrystalline and α-bunn. The polyamide unit cell contains fully extended planner zigzag chains, where adjacent macromolecules are joined by hydrogen bonds of the amide (–CO–NH–). The shifting of the intermolecular hydrogen bonds generates different crystalline forms. The α Polyamide has a planer extended chain conformation. This form is characterized by sheets of anti-parallel chains with the amide group parallel to the rolled plane. The unit cell of α structure is monoclinic with 8 monomeric units and has the dimensions: $a = 9.56$ Å, b (chain axis) $= 17.24$ Å, $c = 8.01$ Å and $\beta = 67.5°$. The space group is P2$_1$, P 2$_1$/m, P2$_1$/c.

The second crystalline form of Polyamide 6 is referred as γ. It is due to a parallel relative shifting of alternate chains in the hydrogen-bonded sheet by about one atom. The unit cell is composed of extended parallel chain segments. The hydrogen bonding is of the pleated sheet rather than the co-planer sheet type. The amide groups lie at about the same level in the cell and the intermolecular chain distance is about 4.8 Å. The unit cell of γ structure is a pseudo hexagonal cell and its dimensions are $a = b = 4.79$ Å, $c = 16.7$ Å, $\alpha = \beta = 90°$ and $\gamma = 60°$. Some other suggests a monoclinic unit cell, whose dimensions are $a = 9.33$ Å, $b = 16.88$ Å, $c = 4.78$ Å and $\alpha = 121°$. The crystal structure has the space group C$^5_{2n}$-P2$_1$/a. The cell contains four monomeric units. The chain repeat distance of the γ form is shorter than that of α form and the difference is independent of the number of carbon atom involved. This results in the rotation of the amide group by about 60° around C–C and C–N bond. The NH $\cdots$ CO hydrogen bond length between the adjacent molecular chains in the parallel sheet is approximately 2.8 Å. The hydrogen density of the γ-form is much less than that of the α form.

The crystallites in Polyamide 6 are made up of hydrogen bonded sheets and the normal method of packing involves a shear of 3/14.b between sheets ($b = 17.2$ Å). The appearance of the hydrogen bonded sheet is shown schematically in Fig. 3.6. The shear is staggered leading to a doubling of the axial length in the sheet attacking direction. The α axis is doubled because of the inversion of successive molecular chains to perfect the hydrogen bonding and the c-axis is doubled because of the staggered rather than progressive shear. In Polyamide 6 the planes of the molecular chains are twisted out of the (001) plane towards the longer diagonal of the projected cell base.

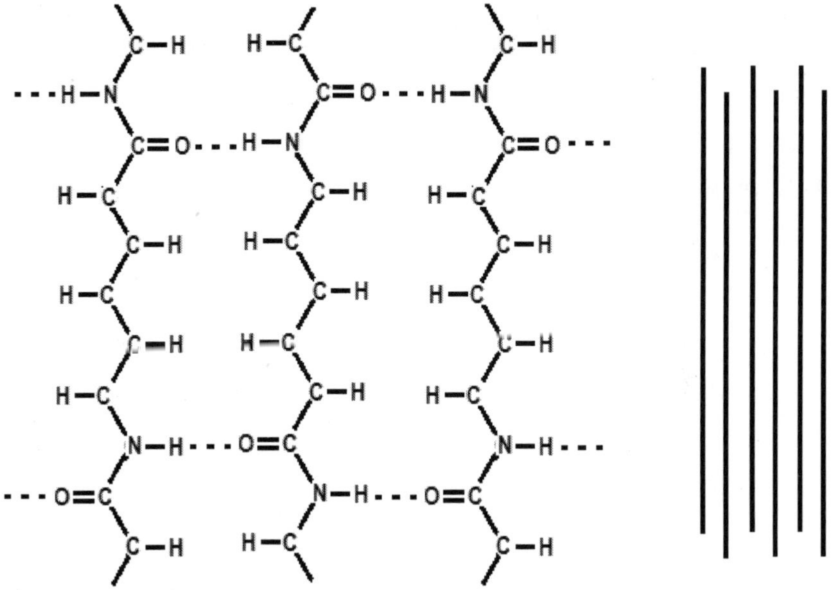

Figure 3.6 H-bonded sheets of Polyamide 6 with the chains in staggered shear effect.

The α form is the thermodynamically most stable crystalline form and can be obtained by slow cooling from the melt. Polyamide 6 crystallizes into the stable α form, only if the initial material is amorphous. The γ form is obtained by spinning fibres at high speed or by iodinating Polyamide 6 in an aqueous potassium iodide and iodine treatment followed by removal of the iodine and potassium iodide with sodium thiosulphate. γ form can be converted to α form by stress. It is a first order crystal transition. The mechanism of the α-γ transition by iodine treatment is a phenomenon of crystal alignment. Iodine enters into the crystal lattice through the space between the hydrogen bonded molecular sheets and co-ordinates to the oxygen of the amide groups lying in the adjacent sheets. Consequently, hydrogen bond is broken and the amide group is preferentially twisted to a direction roughly perpendicular to the molecular sheet instead of coupling with the original one. The molecules are then joined together by new hydrogen bonds and a new hydrogen bond is then formed between the molecular sheets. This molecular sheet is referred as parallel sheets corresponding to the anti-parallel sheet of α Polyamide. γ form does not convert into α form unless the hydrogen bonding is severely affected either by heat treatment or mechanical force or by swelling in selected solvents.

Any small molecular ordering always results in mixtures of α, γ and metastable phases. α and γ phases are interspersed in the lamellae with the metastable phase at the interface. So drawing and drawing variables such as draw ratio, draw temperature and annealing increase the relative amount of α phase. γ form can also be converted to α form by treating with a phenol aqueous solution and the α form obtained from such treatments is again converted into the γ form. This indicates that the arrangement of molecular chain directions in the γ form is not different from that in the α form.

Apart from the two phase structure, i.e. crystalline and amorphous structure, Polyamide 6 is characterized by an intermediate state of order. The intermediate state of order is characterized by insufficient number of hydrogen bonds. The structure is visualized as the chains are parallel over a considerable portion of their length. Each molecule can have any orientation and it need not be related to the orientation of the neighbouring chains.

3.3.6 Crystal structure of Polyamide *n*

Like Polyamide 6, Polyamide 3, Polyamide 4, Polyamide 8, Polyamide 10 and Polyamide 12 show polymorphism with different crystalline modifications. Polyamide 3 crystallizes in a planner zigzag extended chain structure. The unit cell is monoclinic with $a = 9.33$ Å, b (chain axis) $= 4.78$ Å, $c = 8.73$ Å and $\alpha = 60°$. There are four units per cell. Polyamide 3 consists of three modifications. Two of these forms are monoclinic structures. Modification IV is a smetic hexagonal structure and is the stable form near the melting point. Modification IV may be converted into the major monoclinic structure by boiling in water. Polyamide 4 fibre exists in an anti-parallel extended chain structure. It has three crystal modifications. The unit cell contains 8 monomeric units per cell with dimensions $a = 9.29$ Å, b (chain axis) $= 12.24$ Å, $c = 7.97$ Å and $\alpha = 114°$. The β form is more stable structure at elevated temperatures. It may be formed at room temperature by quenching. The β form is believed to have a hexagonal structure and exhibits considerable disorder and diffuses of reflections.

The crystal structure of Polyamide 7 exists in an extended chain coplanar hydrogen bonded sheet structure with anti-parallel chains. The unit structure was triclinic and contains one monomeric unit. The dimensions are $a = 4.9$ Å, $b = 5.4$ Å, $c = 9.85$ Å (chain axis), $\alpha = 49°$, $\beta = 77°$ and $\gamma = 63°$. Polyamide 8 resembles Polyamide 6 exhibiting two crystalline forms closely related to the α and γ forms exhibited by this polymer. Indeed, they differ mainly from those exhibited by Polyamide 6 in the chain axis repeat distance. The four carbon atoms that must be added to the two residues in the repeat unit correspond to 5.1 Å, which represents the major differences. The unit cell dimensions are $a = 9.8$ Å, $b = 22.4$ Å, $c = 8.3$ Å and $\beta = 65°$. The γ form, which is hexagonal, is more stable. Special phenol treatment can be done to obtain to the monoclinic α form

Polyamides 9 and 11 resemble Polyamide 7 and Polyamides 10 and 12 resemble Polyamide 8 and to a lesser extent Polyamide 6. Polyamides 9 and 11 possess triclinic unit cells containing extended chains in anti-parallel configurations in coplanar sheets. Polyamide 10 is polymorphic exhibiting two crystalline forms. The γ form is most stable. Treatment with phenol produces α form. Polyamide 11 is a triclinic unit cell with dimensions: $a = 4.9$ Å, $b = 5.4$ Å, $c = 14.9$ Å, $\alpha = 49°$, $\beta = 77°$ and $\gamma = 63°$. In Polyamide 12, the γ form is also obtained and appears to be more stable than the α form, which has been obtained in single crystals. In Polyamides 8, 10 and 12, the α and γ forms are observed but the γ form is preponderant whereas α form is more preponderant in Polyamides 4 and 6.

3.3.7 Crystal structure of Polyamide 6,6

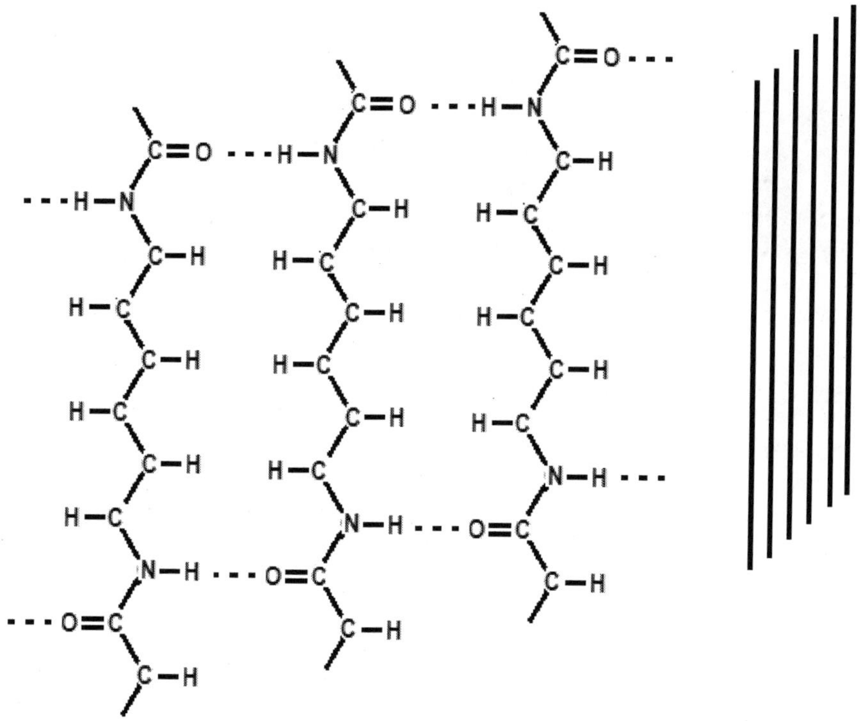

Figure 3.7   H-bonded sheets of Polyamide 6,6 with the chains in
                        progressive shear effect.

Polyamide 6,6 is a triclinic unit cell with dimensions: $a = 4.9$ Å, $b = 5.4$ Å, $c = 17.2$ Å (chain axis), $\alpha = 48.5°$, $\beta = 77°$, and $\gamma = 63.5°$. Polyamide 6,6 molecules exhibit a centre of symmetry and there is no distinction between and anti-parallel chain placement. The unit cell contains only one chemical repeat unit and adjacent hydrogen bonded sheets are displaced parallel to one another in the chain axis direction by the equivalent of three chain atoms in order to accommodate the packing of the amide groups. The appearance of the hydrogen bonded sheet is shown schematically in Fig. 3.7. The β form corresponds to a polymorphic form produced by staggered displacements of the hydrogen bonded sheets in the chain axis direction. The dimensions are: $a = 4.9$ Å, $b = 8.0$ Å, $c = 17.2$ Å (chain axis), $\alpha = 90°$, $\beta = 77°$, and $\gamma = 67°$. As the temperature of Polyamide 6,6 is raised, the distance between the two principal reflections, i.e. the 100 and 010, decreases and eventually coalesces at 175°C. This was first noted by Brill and known as Brill temperature, which corresponds to a transformation from a triclinic to a pseudohexagonal unit cell.

The structure of crystalline portions in Polyamide 6 and Polyamide 66 fibres are very similar. The crystallites are made up of hydrogen bonded sheets and in both the normal method of packing involves a shear of 3/14.b where b is the fibre axis and it is 17.2 Å. The appearance of the sheet is shown in figure. In Polyamide 66 α form, the shear is progressive while in Polyamide β form, the shear is staggered. The planes of polymer chains are twisted towards the shorter diagonal. If Polyamide 6,6 is quenched rapidly from the melt to 0°C, a disordered, mostly nematic paracrystalline material is obtained. This material exhibits diffuse, poorly defined WAXS reflections. If the material is heated above Brill temperature and then slowly cooled the α triclinic structure may be obtained.

3.3.8 Crystal Structure of Polyamide m,n

Apart from Polyamide 6,6, fewer polyamides like Polyamide 6,10, Polyamide 4,6 and Polyamide 7,7 have some interest. Polyamide 4,6 is getting more importance for an alternative fibre instead of Polyamide 6,6. Polyamide 6,10 though commercialized but it is unsuccessful because of inferior properties than Polyamide 6,6. The crystal structures of Polyamide 6,6 and Polyamide 6.10 are completely analogous. Fibres of these two polyamides usually contain two different crystalline forms, α and β, which are different packing of geometrically similar molecules; most fibres consist chiefly of the α form. For Polyamide 6.10, like Polyamide 6,6, the unit cell of the α form is triclinic, $a = 4.95$ Å, $b = 5.4$ Å, c (fibre axis) $= 22.4$ Å, $\alpha = 49°$, $\beta = 76.5°$, $\gamma = 63.5°$. This differs from Polyamide 6,6 essentially only in a 5.2 Å addition to the chain axis repeat distance. This clearly corresponds to four carbon atoms in a planner zigzag conformation. One chain molecule passes through the cell in both cases. The chains are planar or very nearly, so the oxygen atoms appear to lie a little off the plane of the chain. The molecules are linked by hydrogen bonds between CO and NH groups, to form sheets. A simple packing of these sheets of molecules gives the α arrangement. An alternative packing of the sheets gives a two-molecule triclinic cell, and this is the structure proposed for the β form. A β form was detected caused by displacement of hydrogen-bonded sheets along the chain axis. The β-structure is similar to that of α Polyamide 6,6. Polyamide 7,7 exists in an extended chain form that may be represented in terms of hexagonal unit cell. The unit cell parameters are: $a = b = 4.95$ Å, $c = 18.95$ Å (chain axis), $\alpha = \beta = 90°$ and $\gamma = 60°$. Those polyamides, which have γ structures, there is a kink in the backbone, making the chain axis repeat distance less than that computed for an extended zigzag chain.

The reflection ring at 4.14–4.22 Å in x-ray diffractogram for Polyamide 6, Polyamide 6,10 or Polyamide 6,6 is characteristic of polyamides in general. The exact value of this reflection and so the relative strength might be varied with the structure of the Polyamide sample. It can be 4.22 Å for quenched sample and 4.14 Å for the annealed sample. This reflection is due to molecular arrangement and is characteristic of poorly crystalline samples. These observations conclude that the anomalous ring is due to a molecular arrangement and is characteristic of poorly crystalline sample that differs from the recognized polyamide crystal structure. In moderately crystalline sample it can exist together with the usual structure and accordingly such samples can represent a mixture of two structures. The largest perpendicular distance between adjacent chains in the recognized polyamide structure 4.8 Å, i.e. the separation within the hydrogen bonded sheets. This can be presumed that the chains as cylinders of diameter 4.8 Å and these cylinders form parallel closely packed arrays.

3.3.9 Crystal structure of poly (ethylene terephthalate)

Poly (ethylene terephthalate) does not have any active groups. So the interchain interaction is weak. There is certain dipole–dipole interaction and van der Waals force of interaction present between chain molecules. PET molecule is present in a fairly extended conformation. The molecular length of the repeat unit is estimated to be 10.75 Å. There are four repeat units in a unit cell. The crystal structure is a triclinic structure having the following dimensions: $a = 4.56$ Å, $b = 5.94$ Å, $c = 10.75$ Å (chain axis), $\alpha = 98.5°$, $\beta = 118°$ and $\gamma = 112°$. The methylene groups are nearly normal to the chain axis of the molecule. Part of the amorphous region is associated with the extended chain units making up the interlamellar links although it is relatively small. In amorphous state PET has a gauche structure of ethylene glycol fragment contrary to the crystalline state where the ethylene glycol fragment of PET has a trans structure (Fig. 3.8). All major changes, which are observed upon transition from the crystalline to the amorphous state, have been attributed to this conformational change. The as spun fibre is mostly amorphous. After drawing, the stress induced crystallization results in formation of crystalline structure of poly (ethylene terephthalate). Annealing at high temperatures of this fibre leads to the formation of large lamellar structures. Poly (butylene terephthalate) is another polymer used as a fibre. This polymer consists of lower aromaticity than that of poly (ethylene terephthalate). Like poly (ethylene terephthalate), poly (butylene terephthalate) consists of a triclinic structure. The dimensions are: $a = 4.83$ Å, $b = 5.94$ Å, $c = 11.5$ Å, $\alpha = 99.7°$, $\beta = 115.2°$ and $\gamma = 110.8°$. This polymer also shows another modification, i.e. β modification. The pure β form can be obtained by stretching pure α form about 12% or more and the β form is transformed spontaneously to pure α form when the stress is removed.

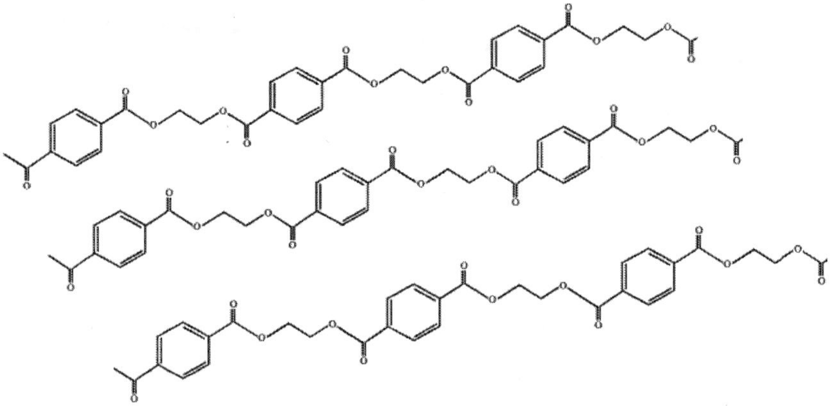

Figure 3.8   Molecular structure of poly (ethylene terephthalate).

3.3.10 Crystal structure of polyacrylonitrile

Polyacrylonitrile fibre has a repeat unit of acrylonitrile, i.e. $CH_2–CH–(C\equiv N)$. The chemical structure, i.e. chain molecule of polyacrylonitrile is shown in Fig. 3.9(a). The chain is fully repetitive and contains $C\equiv N$ groups which have a triple bond and therefore very active. Owing to the asymmetrical presence of CN group, the chain molecule exhibits tacticity, i.e. syndiotactic structure as well as isotactic structure. Due to this bulky group, the configuration of polyacrylonitrile molecule is in helical form with a syndiotactic structure as shown in Fig. 3.9(b). The molecule of polyacrylonitrile is often described as a rod-like molecule, like that of (c) in Fig. 3.9. The nitrile groups are distributed around the rod surface at different angles to the axis. These rod-like molecules attract each other strongly because of electrical dipolar forces. So the molecules pack together with varying degrees of lateral order, but with little longitudinal order along the fibre axis. It is interesting to note that polyacrylonitrile fibre heated above 80°C pass through a transition stage and the molecular chain shows to shrink. This also induces a pasticization effect on the structure. The chains distorted, bend, twist in the fibre structure and orientation is lost. Another interesting observation for PAN fibre is the presence of microvoids. The porous zones are spatially distributed. Thermal relaxation of intercrystalline region leads to an increase in its porosity.

There is no cross linkages in the chemical structure of poly acrylonitrile. But there are strong secondary valency forces between adjacent chains due to the formation of hydrogen bonds, existing between the α hydrogen atom of one chain and the nitrile nitrogen of an adjacent chain for which it is insoluble. A distance of 5.3 Å separates the chain, the distance between the polymer chains through the nitrile group. The chain molecules tend to be bonded together to form aggregates as a result of active CN group. These aggregates are called 'fibrils' and the formation of these fibrils takes place during the formation of fibres in spinning.

The fibre exhibits a type of crystalline order extending laterally from one molecular chain to another but little or none along the chain axis. Total degree of crystallinity in the fibre is only moderate. However, infrared spectroscopic investigation indicates that PAN is neither amorphous nor crystalline but it is mostly mesomorphic. Although PAN shows thermoplasticity at comparatively lower temperature, but when heated to a higher temperature a ladder polymer consisting of six-membered ring is formed owing to the linkage between the groups, as shown in Fig. 3.10. The ladder polymer is more stable towards heat than the original chain and does not melt easily.

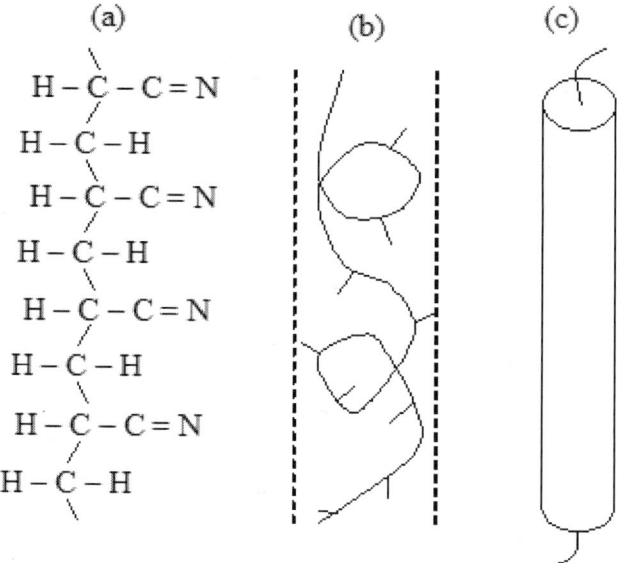

Figure 3.9   Structure of acrylic fibre (a) chemical structure (chain molecule), (b) helical structure, (c) rod-like structure.

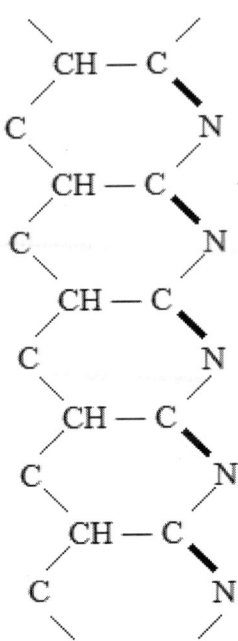

Figure 3.10   Formation of ladder structure of acrylic fibre.

Further readings

1. J. E. Mark (Ed.), *Physical Properties of Polymer Handbook*, Springer, 2007.

2. L. H. Sperling, *Introduction to Physical Polymer Science* (2nd ed.), Wiley, New York, 1992.

3. Robert J. Young and Peter A. Lovell, *Introduction to Polymers*, CRC Press, 2011.

4. J. W. Hearle and R. H. Peters (Ed.), *Fibre Structure*, Butterworth, London, 1963, p 346.

5. Claudio De Rosa and Finizia Auriemma, *Crystals and Crystallinity in Polymers: Diffraction Analysis of Ordered and Disordered Crystals*, Wiley, 2013.

6. *Encyclopedia of Polymer Science and Engineering*, Wiley, New York, 1986.

7. M. Lewin and E. M. Pearce, *Handbook of Fibre Science & Technology*, Marcel Dekker, New York, 1985.

8. S. Eichhorn, J.W. S. Hearle, M. Jaffe, and T. Kikutani, *Handbook of Textile Fibre Structure*, Woodhead Publishing Series in Textiles, 2009.

Amorphous structure

4.1 Introduction

The disordered structure of the molecular chains in the fibre and/or polymer is termed as 'amorphous state' or 'amorphous region'. This structure Is scantily ordered and consists of random tangle of molecular chains. The order of the molecular chain length much less than that the chain lengths are arranged in bundles. The intermolecular distances are in the range of 10–20 Å. So it can be assumed that the region consists of minimum interchain forces. The state of disorder is considered to resemble that in a low molecular weight liquid. These regions can undergo a liquid-to-glass transition on cooling and vice versa. The radius of gyration values in the amorphous state is in close agreement with the value for the isolated unperturbed chain. Any local ordering would have been changed the value of the radius of gyration. In general, the radius of gyration value of a chain in ordered state is more or less similar to that of the chain in the stretched position. The amorphous regions are complimentary to the crystalline regions.

The random order of the molecular chains in the amorphous state may be present as (a) cilia, (b) bridges, (c) loops or (d) free chains. Cilia are slender protuberances that project from longer body. Bridges are the chins which reach across the structure. Loop is a structure, the ends of which are connected to the beginning. Free end indicates that the whole chain is not fixed nor have any bindings. This means that the randomness of the molecular chains differ in different state of the structure. The amorphous region in between the crystallites consists of both bridges and loops. The molecular chain length of a polymer of the fibre is dependent upon the molecular weight and degree of polymerization and it is in the range of 500–1500 Å. For the same polymer of the fibre, the crystallite length and the amorphous region length can be approximately, 80–150 Å and 20–60 Å, respectively, depending upon the crystallinity. So a molecule is usually traversing only about 1/10th of the plane surface area between the two crystalline regions. This means that the chain will have an approximate random walk in the amorphous region.

The loops are not always possible for adjacent re-entry. Based on this, two divergent viewpoints are present related to the state of amorphous region. Those are: (a) coil model and (b) bundle model and these are discussed below:

(a) **Coil model**: The coil model is assumed that the amorphous chains are virtually devoid of all vestiges of order, even at a level approaching the diameter of the chain. The material is homogeneous structure and that the configuration statistics of a single molecule in the melt or in the glassy state is same as that of an unperturbed molecule in solution. The molecular conformation in the bulk and in the solvent should be identical.

(b) **Bundle model**: Bundle model is based on the assumption of domains with nematic liquid crystal like arrangements of the macromolecule. There is existence of some local organization in the melt and it shows anisotropy. The polymeric chains in the amorphous region are organized in small bundles, nodules, meander arrays or paracrystals. The sizes are generally believed to be in the range of 50–100 Å. In a strict sense, the system is not amorphous.

Most polymers are either completely amorphous or have an amorphous like component even if they are crystalline. Major commercial fibres like polyester, nylon and polypropylene contain approximately 40–50% amorphous regions in their structure. The structure of polyethylene fibre is highly crystalline. On the other hand, polystyrene, polyvinyl chloride and polymethyl methacrylate (PMMA) polymers are amorphous polymers. The amorphous polymers do not melt at a fixed melting point with the absorption of a characteristic latent heat but rather soften over a quite range of temperature centred about a mean value, i.e. glass transition temperature (T_g).

Table 4.1   Degree of order in different structures

Type	Degree of order
Extended chain	High
Folded chain	High
Glassy state	Disorder, possibly local order
Rubbery state	Entirely disorder

Qualitatively, the molecular motion is greatly restricted as the polymer or the fibre is cooled from melt to the glassy state through the rubbery state. The transition from the rubbery to the glassy state is usually achieved at glass transition temperature. The solid-state structure related to the amorphous structure consists of two extreme propositions and it is accounted for by bond rotation, bond deformation or bond stretching or by movement of the molecules in an intermolecular force field. One is that of rubbers with minimum or very little intermolecular interactions occur between the chains. The other is that of fully extended molecular chain and or it can be the folded chain crystal. The glassy state of the amorphous structure may be described in terms of aggregate of unit crystals of extended chains with a very small degree of local order between chains. The local order is persisting over a few repeat units along with perpendicular to the chains. The difference arises from the respective natures of the bonding forces involved, i.e. covalent forces along the chains and the van der Waals or secondary forces perpendicular to them. The degree of order will differ in different structures and these are shown in Table 4.1.

If there will be bulky groups or side groups or branches, a close packing hexagonal arrangement of the chain axes might be possible provided these groups can be packed in the direction of the main polymeric chains and do not mutually interfere with each other. Of course, there will be no mutual interference if the groups in each polymeric chain are at different levels. If these side groups will be large or more bulky or cannot be suitably packed in parallel position, a loose packed or open structure will be expected.

4.2 Amorphous solids

Amorphous solids were considered to be disordered crystalline solids, thus originating from the corresponding crystalline state. One could say that to be disordered, there should exist a prior ordered state as a reference from which disorder can be measured. However, many amorphous solids do not have a corresponding crystalline form. Yet, it is still frequently assumed that amorphous materials are found at the limits of disorder, and amorphousness is usually defined by what it is not, rather than by what it is. A survey of experimental techniques will lead one to three methods for positively identifying amorphous materials: (i) calorimetry to measure glass transition temperature or associated thermodynamic quantities; (ii) x-ray scattering yielding the amorphous halo and (iii) NMR to determine bond correlation. The last two methods essentially measure the same structural quantities. All of amorphous solids (natural and synthetic) can be separated into two broad categories: (a) Inherently non-crystallizable and (b) Crystallizable.

The first category is cross-linked polymers or atactic organic polymers. The second category includes metallic and metalloid glasses, some metal oxides and linear polymers in which crystallinity can be developed by annealing. Both characteristics can be combined, as in semi-crystalline polymers. All of these solids display common amorphous characteristics, such as x-ray scattering patterns in the form of an amorphous halo and the presence of glass transition temperature. From the point of view of chemistry, all of the known amorphous materials can be segregated into (at least) 4 cognate classes as listed below.

(1) Metallic glasses: A relatively new group of solids that solidify in amorphous atomic structures are metallic glasses of the metastable category. Such man-made metallic glasses are heterogeneous in composition on atomic scale. X-ray or electron scattering patterns show broad peak (amorphous halo) usually attributed to amorphous solids.

(2) Inorganic glasses: These are strong glasses, where annealing above the glass transition temperature may lead to partial crystallization.

(3) Organic glasses: These glasses naturally divide into three subclasses:

   (a) random 3-dimensional (cross-linked) networks of copolymers,

   (b) polymeric glasses based on atactic (non-cross-linked) macromolecules,

   (c) polymeric glasses based on linear (iso- or syndio-tactic) macromolecules.

   In the first subclass random networks form on copolymerization of wide range of molecules with functionalities greater than 2.0. The cross-linking reaction between the components entrenches spatially random bonding resulting in a random, permanent, non-reversible molecular network. The second subclass includes branched polymers with atactic architecture of the macromolecular chain. These macromolecular solids are restricted to form glasses on cooling from the molten state because the irregular chain structure prevents self-ordering and registering of the monomer units. The polymeric materials have permanent covalent bonds, unchanging between liquid and solid state. Annealing above the glass transition temperature does not lead to crystallization. The third subclass includes linear polymers that can be vitrified by quenching. On annealing, these glasses will develop crystallinity.

(4) Amorphous thin films: Materials produced by methods akin to molecular vapour deposition techniques can form amorphous layers. These materials lack crystalline areas. On heating, the amorphous structure rearranges and the transformation is non-reversible.

4.3 Glass transition temperature

Glass transition temperature (T_g) of a polymer or fibre is defined as the temperature at which a fibre or polymer changes from the rubbery state to the glassy state on cooling or from the glassy state to the rubbery state on heating. Glass transition temperature is a secondary transition. The molecules below this temperature are glass like with minimum molecular mobility. Above glass transition temperature, large segments of the molecule become more mobile. Basically glass transition is the passage of a mobile liquid into the solid state with no change in phase. When a liquid is cooled, its viscosity increases and its energy of thermal motion decreases. This prevents the molecules from regrouping. When the material assumes a glassy state, (below T_g) its properties change. It loses the properties characteristic of the liquid state and acquires the properties of a solid. These changes do not occur suddenly, but gradually over a certain temperature interval of about 10°C–20°C. Owing to this, glass transition temperature is not a point, but it is average of the interval.

T_g values of some of the polymers are given in Table 4.2. At room temperature, rubber is soft and elastic but when cooled to a temperature to approximately −70°C, it assumes hard, brittle and glassy characteristics. At temperature above glass transition temperature, the amorphous polymer is soft and flexible and is either an elastomeric or a very viscous liquid. In semi-crystalline polymers or fibres, the effect of glass transition is less since it occurs only in the non-crystalline parts of the polymer. The molecule of non-crystalline regions is dispersed in an irregular conformation, each molecule having a randomly kinked form. Below T_g, i.e. in the glassy region, the thermal energy is insufficient to surmount the potential barriers for translational and rotational motions of segments of the polymer molecules. So the chain molecules are frozen in fixed positions. Owing to this, the molecules have a high resistance against being deformed because of the strong intermolecular force of attraction present. So these materials are hard, rigid glasses below glass transition temperature. When heated above T_g, the molecules attain a greater degree of flexibility due to the reduction of the intermolecular forces and thus the fibre or polymer structure is less rigid, more flexible and soft.

Table 4.2   Glass transition temperature of selected polymers

No.	Polymer	T_g (°C)
1	Natural rubber	−75
2	Poly isobutylene	−70
3	Polyethylene	−60
4	Polychloroprene	−50
5	Polyvinyl fluoride	−40
6	Poly-1-butene	−25
7	Polyvinylidene fluoride	−20
8	Polypropylene (iso)	−10

9	Polyvinyl acetate	+30
10	Polyvinyl chloride	+85
11	Polyvinyl alcohol	+85
12	Polyamide 6	+40
13	Polyamide 6,10	+40
14	Polyamide 6,6	+57
15	Polyethylene adipate	−70
16	Polybutylene terephthlate	+40
17	Polyethylene terephthlate	+69
18	PMMA	+45
19	Polyacrylonitrile	+90
20	Polystyrene	+100
21	PTFE	+126
22	Polycarbonate	+150

On the molecular state, glass transition temperature may be described as the temperature corresponding to the onset of major segmental mobility in a heating process. At T_g, the micro-Brownian motion of the molecular chain initiated. The polymer acquires short-range diffusional motion along with micro-Brownian motion. Segments are free to move from one lattice site to another and the hard polymer becomes soft and rubbery. The molecular mobility or the conformational rearrangement leads to the flexible behaviour of the polymer or the fibre. The intermolecular cross-links break down, increasing mobility of the units and the flexibility of the chains and the polymer passes into high elastic state.

Glass transition temperature depends upon the ratio between energy of interaction and the energy of thermal motion of their units. The energy of intermolecular interaction is independent of temperature. But the energy of thermal motion decreases sharply with decreasing temperature. At certain definite temperatures, energy of thermal motion is insufficient to overcome the energy of interaction as a result of which chain rigidity increases. The chain rigidity increases upon cooling. On the other hand, cooling results in the formation of a stable structure with a fixed random arrangement of polymer molecules relative to one another.

The glass transition temperature depends on the structure and polarity of the polymer, which affects both chain flexibility and intermolecular interaction energy. T_g depends on the restrictions on molecular mobility. The greater the restriction on mobility, the higher will be the T_g. The restrictions on molecular mobility can be varied because of the changes in structure and/or external factors. The structural variables that affect glass transition temperature are considered on the basis that the thermal energy supplied to the polymer overcomes (a) cohesive forces holding the component together and (b) inherent flexibility of the chain segments. Non-polar polymers possess high chain flexibility and their chains remain flexible down to very low temperature. The energy of interaction between polar groups is several times higher than the non-polar groups. A polymer with higher polar groups exhibits high glass transition temperature and vice versa. Presence of any substituent polar group or bulky groups hinders rotation of the units and the chains will have reduced flexibility and higher glass transition temperature.

Glass transition temperature is basically a property of amorphous state. So only the amorphous polymers or semi-crystalline polymers or fibres containing a substantial amount of amorphous region show glass transition temperature. The restrictions on molecular mobility can be varied because of the changes in structure and/or external factors. Each polymer or fibre has its own characteristics T_g because of the factors like (1) Chemical structure, (2) Degree of polymerization and its distribution (molecular weight and molecular weight distribution), (3) Symmetry of the structure, (4) Molecular polarity, (5) Steric hindrance, (6) Flexibility of the molecular chain, (7) Cross linking, (8) Bulkiness of the side group attached to the backbone chain, (9) Blocking or grafting or polymer blending.

The glass transition temperature is of practical importance as it divides the polymer or the fibre into those used in relatively brittle, glassy state and those used in their rubbery state. The material is not useful if its transition temperature is at room temperature or near room temperature. Further deformation or any chemical treatment is difficult in glassy state, because of the rigidity of the molecular chains. On the other hand, it can be possible because of the co-operative motion of local segments and flexibility of the chain segments. This can be achieved if the polymer or the fibre is in a rubbery state, i.e. above glass transition temperature.

4.4 Theories of glass transition

Glass transition can be analysed and described by means of different theories. There are at least three major theories, which analysed the concept and theories of the glass transition. Those are: (1) Free Volume Theory, (2) Barrier or Relaxation Theory and (3) Statistical Theory.

4.4.1 Free volume theory

Free volume theory is closely related to the hole theory of liquids. The total volume per mole ϑ is treated as the sum of the free volume 'ϑ_f' and an occupied volume 'ϑ_o'. Free volume includes van der Waals radii and the volume associated with vibrational motion. So free volume is the extra volume required for large-scale vibrational motions for those found between consecutive atoms of the same chain. Flexing over several atoms that is transverse string like vibrations of a chain rather than longitudinal or rotational vibrations will obviously require an extra room.

It is assumed that the temperature co-efficient of expansion of the free volume is greater than that of the occupied volume above T_g. T_g is defined on the free volume concept as that temperature at which ϑ_f collapses sensibly to a minimum value, or at any rate to a frozen in value as shown in Fig. 4.1. Hole mobility has therefore been totally restricted and the only movement below T_g is that allowed by the occupied volume ϑ_o. Thermal expansion of a liquid without change of phase can be regarded as an increase in free volume. If the specific volume at temperature T_o as ϑ_{T_o}, then the specific volume at temperature T will be

$$\vartheta_o = \vartheta_{T_o} \{1 + \alpha_T(T - T_o)\} \tag{4.1}$$

where α_T is the co-efficient of thermal expansion at temperature T. So the fractional increase of volume f_ϑ will be

$$f_\vartheta = (\vartheta_o - \vartheta_{T_o})/\vartheta_o = \alpha_T T \tag{4.2}$$

The specific volume of free space or free volume will be

$$\vartheta f = \vartheta T - \vartheta_{T_o} \tag{4.3}$$

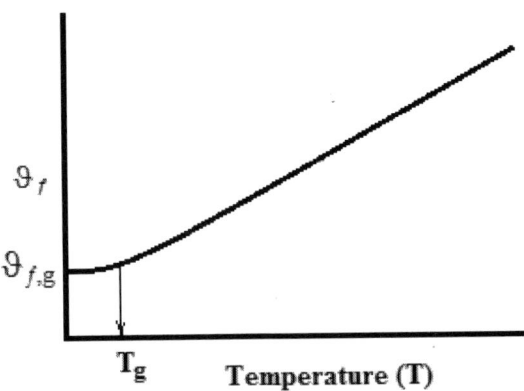

Figure 4.1   Relationship between free volume (ϑ_f) and temperature (T).

Equations (4.1) and (4.3) can be modified with the concept of free volume at glass transition temperature as shown in Fig. 4.1 and Eqs. (4.4) and (4.5) as the amount of free volume below glass transition temperature is practically constant and it can also be neglected.

$$\vartheta_f = \vartheta_{T,g}\left\{1 + \alpha_T\left(T - T_g\right)\right\} \tag{4.4}$$

$$\vartheta_f = \vartheta_T - \vartheta_{T,g} = A \cdot T^{3/2} \tag{4.5}$$

where A is constant

For most polymeric materials like fibres, the fraction of free volume at glass transition temperature ($\vartheta_{f,g}$) in proportionately with specific volume at T_g ($\vartheta_{T,g}$) is approximately 0.0025 ± 0.0003, i.e.

$$f_g = \vartheta_{f,g}/\vartheta_{T,g} = 0.0025 \pm 0.0003$$

Equation (4.4) can be modified as

$$f_g = 0.0025 + \alpha_T\left(T - T_g\right)$$

4.4.2 Barrier theory

The molecular interpretation of the glass temperature can be done by means of the barrier theory based on free rotation of the atomic groups. This is referred as barrier theory. This theory uses a temperature dependence viscosity term, identified as 'friction factor' or resistance to movement of chains either parallel to each other or rotational. The resistance in turn can be due to potential barriers (Fig. 4.2). It can be assumed that the temperature affects the response time of the system to external change. An increase in temperature causes the response time to be reduced, and at these conditions the rapid deformation will be responded to as if they were slow and the material is rubbery. Conversely a fall in temperature causing a sluggish response will produce the effect of grass.

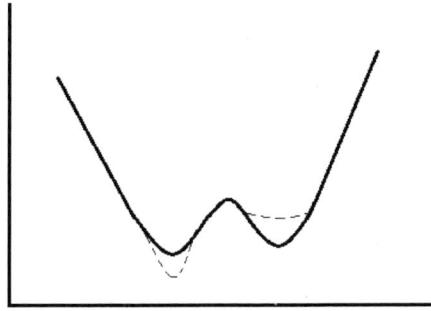

Figure 4.2   The potential barriers in a molecule.

At equilibrium, there is equal probability from one state to the other.

$$P_{21} = P_{12} = A \exp [-\Delta H/RT] \qquad (4.6)$$

where ΔH = activation energy and it is assumed $\Delta H \gg RT$ and A is a constant.

If mechanical or other external conditions applied so that one position becomes more favourable than the other, then the situation will be modified and it is shown as per the dotted lines (Fig. 4.2). One will be lowered by $\vartheta/2$ and the other will be raised by a similar amount.

Now
$$P_{12} = A \exp [-(\Delta H + \vartheta)/RT] \qquad (4.7)$$

$$P_{21} = A \exp [-(\Delta H - \vartheta)/RT] \qquad (4.8)$$

And it is assumed $\vartheta \ll RT$

Let
$$P = A \exp [-\Delta H/RT] \qquad (4.9)$$

then
$$P_{12} = P[1 - \vartheta/RT] \qquad (4.10)$$

$$P_{21} = P [1 + \vartheta/RT] \qquad (4.11)$$

If the wells 1 and 2 are occupied at any time be N_1 and N_2 links, respectively, where $N_1 + N_2 = N$, then flux will be

$$dN_1/dT = -N_1 P_{12} + N_2 P_{21} \qquad (4.12)$$

and
$$dN_2/dT = -dN_1/dT \qquad (4.13)$$

Hence with incorporation of Eqs. (4.10) and (4.11), in Eqs. (4.12) and (4.13), one can obtain

$$d(N_1 - N_2)/dt = 2P(N_1 - N_2) + 2P(N_1 + N_2)(\vartheta/RT) \qquad (4.14)$$

At equilibrium
$$N_1 P_{12} = N_2 P_{21} \qquad (4.15)$$

The approach to equilibrium is given by

$$N_1 - N_2 = N\vartheta/RT (1 - e^{-2Pt}) \qquad (4.16)$$

The response time or relaxation time $\tau = (2P)^{-1}$ \qquad (4.17)

$$\tau = (2P)^{-1} = (1/2A) \exp (\Delta H/RT) \qquad (4.18)$$

Temperature effects may be explained by means of relaxation time τ. Secondary transitions often do behave similar way and the energy barrier interpretation of relaxation is often applied to them with consequent interpretation of the activation energy in molecular terms. For multiple relaxation times, there will be multiple barrier models for analysis of those relaxation phenomena.

4.4.3 Statistical theory

There are statistical treatments, which aim at interpreting the phenomenon in terms of molecular process. When enthalpy (ΔH/Cal g^{-1}) or entropy or specific volume plotted with temperature, an apparent paradox is usually observed near glass transition temperature. The enthalpy of the liquid phase is rapidly approaching equality to that of the crystalline phase, when vitrification sets in. Similar type of observation can also be observed in case of entropy or specific volume. The extrapolation of the liquid phase below T_m would reach at a temperature T_x between absolute zero and T_m. At this temperature, the material would either crystallize or exist in a state of lower entropy than the crystalline state and it is impossible. Thus T_x would be an equilibrium transition temperature in the true thermodynamic sense, contrary to the assumption that the glass transition was not a true thermodynamic transition.

When the entropy between liquid and crystal can be compared, a glass could be obtained with entropy less than that of the crystals. A lattice model related to the entropy of the polymer is generally used for statistical calculation of the conformational entropy. By making the assumption that the one orientation of a link in a chain relative to its neighbour is energetically preferred over all others, then a temperature exists at which conformational entropy vanishes, giving a true second order transition at that point. The number of available states in the statistical mechanical sense is not constant through the glass transition but reduces to zero at some temperature theoretically, which may never be reached in an experimentally realizable time.

The statistical theory concentrates entirely on the barriers to rotation and this theory neglected intermolecular cohesion. This theory may not be applicable to low molecular weight polymers, as these polymers can be present in glassy state. This T_x should be considered as a theoretically derived equilibrium glass transition temperature, which would in theory be attained by an infinitely slow cooling from melt. This temperature will depend on the chain rigidity and on the energy of intermolecular interaction. Physically the process of cooling a polymer through its glass transition would modify the material in different ways. At high temperature, the configuration entropy will be high and there will be many ways for molecules to be packed with no one configuration-preferred over another. As the temperature falls, however, two processes occur: (1) Low energy configuration starts to predominate and (2) the volume of holes decreases. The number of ways of packing is reduced since the empty volume available becomes more and more correlated with the molecules. Eventually, the limit is reached where no further packing would be possible and where the conformational entropy therefore vanishes. This is T_x, and it is proposed as second order transition, as per statistical theory.

4.5 Chemical structure and T_g

Several factors, which can be related to chemical structure, can affect T_g. Those are chain stiffness or flexibility, cohesive energy density, solubility parameter, backbone symmetry, the barriers hindering rotation around the bonds, the free volume and the number of links in the chain. The flexibility and cohesive energy density or polarity of each group is nearly independent of the other groups in the molecule to which they are attached. Because of this, each group can be assigned an apparent T_{gi}, and the T_g of a polymer becomes the sum of the contributions of all the groups, i.e.

$$T_g = \Sigma T_{gi} \cdot n_i \qquad (4.19)$$

where n_i is the mole fraction of ith group in the polymer. So any change in apparent glass transition temperature of the group will also change the ultimate glass transition temperature of the polymer.

(a) Flexibility

The most important factor is chain stiffness or flexibility of the polymer molecular chain. Long chain aliphatic groups or linkages present in the molecular structure induce flexibility and lower T_g value of the polymer. On the one hand, the molecular chains can be more rigid if aromatic structures and pendant tertiary butyl groups can be substituted. Presence of these relatively large and rigid groups raises the glass transition temperature. The lowering of molecular flexibility by the substitution of bulky side groups on to a polymer reduces the flexibility and affects T_g. However, it is the flexibility of the group, not its size that is the major factor determining T_g. For example, the T_g of polyethylene (−120°C), polypropylene (−10°C), polystyrene (+100°C) and poly (2,6 dichloro styrene) (+167°C), where T_g of polystyrene and poly (2,6 dichloro styrene) became much higher due to presence of aromatic groups in the structure.

(b) Molecular polarity

Molecular polarity or cohesive energy density of the polymer influences the value of T_g. Cohesive energy density (CED) or solubility parameter (δ), best describes the energy of interaction. CED is defined as the energy of vapourization (E_v) divided by its molar volume. This measures the cohesion per unit volume of liquid. This measures the cohesion per unit volume of liquid.

$$CED = (E_v/v) \tag{4.20}$$

or
$$\delta = (E_v/v)^{1/2} \tag{4.21}$$

Polymers with high solubility parameter (δ) tend to have lower T_g. Polymers with small rotational barriers tend to have low T_g. Polyethylene has a low solubility parameter and a barrier of rotation of 3.3 kcal/mol whereas PTFE ($T_g > 20°C$) have also a low δ_p but a high rotational barrier of 4.7 kcal/mol. Increasing the polarity of the molecule increases its T_g. Thus in the series, polypropylene ($T_g = -10°C$), polyvinyl chloride ($T_g = 85°C$) and polyacrylonitrile ($T_g = 101°C$), the size of the side group is about the same, but the polarity increases. In these three polymers, the glass transition temperature differs because of the polarity of the groups.

(c)   Backbone symmetry

Backbone symmetry affects the shape of the potential wells for bond rotation. Owing to this, the symmetrical polymers have lower glass transition temperature than the unsymmetrical ones. The glass transition temperature of symmetrical polymers like polyisobutylene ($T_g = -70°C$) or polyvinylidene chloride ($T_g = -19°C$) is much lower than asymmetrical polymers like polypropylene or polyvinyl chloride ($T_g = 85°C$).

4.6   Structural factors and T_g

Glass transition temperature can also be influenced by the molecular structure of the polymer and/or the fibre. The molecular structure of a long chain polymeric fibre is characterized by molecular weight, presence of co-monomer (copolymer), plasticization, crystallinity and orientation.

(a)   Molecular weight

The glass transition temperature increases with number average molecular weight $\overline{M}_n$. In practical range of molecular weight, T_g is given by

$$T_g = T^0_g - K/\overline{M}_n \tag{4.22}$$

where T^0_g is glass transition temperature for infinite molecular weight and K, characteristic constant. The change in T_g arises from the ends of the polymer chains, which have more free volume than the same number of atoms in the middle of chains. Molecular weight is related to the molecular chain length and entanglement. Higher the molecular weight higher will be the entanglement and it reduces the free volume. For two different molecular weight polymers (designated as a and b), Eqs. (3.22) can be rewritten as Eq. (4.23)

$$T_{g,a} + K/\bar{M}_{n,a} = T_{g,b} + K/\bar{M}_{n,b} \qquad (4.23)$$

(b) Cross-linking

A cross-linked polymer is characterized by presence of covalent bonds in perpendicular direction. A higher degree of cross-linking leads to formation of network polymers. This restricts the movement or the slippage of the molecular chains. So cross-linking increases the glass transition temperature by the restrictions imposed on the molecular motions of a chain. Low degree of cross-linking increases T_g only slightly above that of the uncross-linked or linear polymer. However in highly cross-linking materials such as phenol formaldehyde resin and epoxy resin, T_g is markedly increased by cross-linking. Mathematically, this can be expressed in Eq. (4.24).

$$T_g - T_g^0 = 3.9 \times 10^4/M_c \qquad (4.24)$$

where M_c is the number average molecular weight between cross linking points M_c, T_g^0 is the glass transition temperature of the uncross-linked polymer having the same chemical composition. $T_g - T_g^0$ is the shift in T_g due to cross-linking after correcting any copolymer effect of the cross linking agent.

(c) Plasticization

Plasticizers are low molecular weight liquids, which when penetrate inside the structure; interact with the structure of the polymer or the fibres and lower glass transition temperature. If T_g of the two component A and B are known, an estimate can be made of the T_g of the mixture by any of Eq. (4.25) or (4.26).

$$T_g = T_{g,A} \cdot \varphi_A + T_{g,B} \cdot \varphi_B \qquad (4.25)$$

or $\qquad (T_g)^{-1} = w_A \cdot (T_{g,A})^{-1} + w_B \cdot (T_{g,B})^{-1} \qquad (4.26)$

where φ the volume fraction and w is the weight fraction of the components. Eqs. (3.25) and (3.26) indicate that the glass transition temperature value of the material can be lowered by means of any solvent. The extent of reduction is dependent upon the glass transition temperature of the solvent as well as the amount of the solvent added to the material.

(d) Copolymer

The glass transition temperatures of copolymers are very analogous to those of plasticized material if the comonomers B is considered to be plasticizer for homopolymer. For the estimation of the resultant glass transition temperature of the copolymer, Eqs. (4.25) and (4.26) can be used. The glass transition temperature of the individual component and their fraction present in the copolymer influences the resultant glass transition temperature of the copolymer.

(e) Structure

The glass transition temperature is a property of the amorphous structure, where minimum amount of intermolecular forces are present between molecular chains in a non-uniform manner. This helps for the chain slippage at a lower temperature and ensures the operation of micro-Brownian motion at a lower temperature. The intermolecular forces are more uniform and higher for crystalline structure. So the glass transition temperature may affect because of the structural modification in terms of crystallinity and orientation. In general, increase in crystallinity increases intermolecular forces, restricts the slippage of molecular chains and reduces the amount of free volume present in the fibre. This will enhance the glass transition temperature. The polymer of the fibre with higher crystallinity may have higher glass transition temperature and vice versa. Presence of intermolecular forces also acts as weak cross-links between molecular chains.

4.7 Measurement of T_g

Glass transition is thermodynamically treated as a second order transition. Owing to this, the first derivatives of enthalpy and volume show abrupt change at the T_g. This principle is used for the measurement of T_g. The methods for determining T_g are based on measurement of specific volume or use of thermal analysis technique like differential scanning calorimetry (DSC), differential thermal analysis (DTA) or dynamic mechanical analysis. Dielectric relaxation techniques are also used for determination of T_g. Measurement of the elastic modulus as a function of temperature over glass transition temperature shows a change by at least an order of magnitude. The modulus of fibres, like polyamide or polyester changes its modulus from about 10^{10} N/m² in glassy state to 10^6 N/m² in rubbery state. In rubbers and even in thermoset polymers like epoxy resins, the modulus changes from about 10^9 N/m² in glassy state to 10^7 N/m² in rubbery state. Similar changes may occur in the dielectric constant, refractive index, density and specific heat. However, there is no latent heat involved and for this reason, the glass transition is sometimes referred as a second order transition. In principle, any physical property, which changes at T_g, can be used for its determination. Some of the common methods are given below.

4.7.1 Specific volume method

The volume change as a function of temperature is measured on a dilatometer, and the result obtained can be plotted for determination of T_g. The specific volume changes linearly with the temperature up to a transition region where a change of slope occurs after which the curve continues linearly but at a steeper gradient. T_g is usually defined as the point at which the tangents of the two curves intersect. The point at which the slope (i.e. the first derivative dV/dT) changes is taken as T_g.

4.7.2 DSC or DTA

In DSC or DTA, the measured quantity is proportional to the rate of heat flow, which is the first derivative of the thermodynamic quantity, either heat or power. Hence the experimental recording of heat flow rate with temperature at constant heating rate gives an abrupt change at glass transition temperature.

4.7.3 Relaxation method

In relaxation method, the quantities measured are storage modulus G', loss modulus G'' and loss tangent tan δ for dynamic relaxation (and dielectric constant E', dielectric loss factor E'' and dissipation factor tan δ, in case of dielectric relaxation). The variations of these quantities for glass transition relaxation are shown in Fig. 6. The peak in G'' or point of inflexion G' (which coincides) is taken to represent the relaxation temperature. The peak in tan δ is at a slightly higher temperature and is also taken to represent the temperature of relaxation T_r. The T_r is related to T_g through a function of frequency of measurement. However, for many practical purposes T_r is taken as T_g, or something close to T_g.

4.7.4 Chemical method

Glass transition temperature can be measured by chemical methods, as this is a property of the amorphous region and this region is more susceptible to the chemicals, plasticizers, solvents or dyes. The penetration of these additives can be assessed either by (a) diffusion of the component, if it is chemically inert with the substrate, or (b) critical dissolution time, if the additives interact chemically with the substrate. The principle is that the additives are fast for diffusion or interaction in the amorphous structure and it is faster, when the substrate is rubbery, i.e. at a temperature higher than the glass transition temperature. However, these methods are not accurate to assess the exact temperature.

4.8　Molecular relaxation

There are different types of molecular motion present in a polymer system. These include side group rotation, main chain rotation and translation. Long chain compounds are crystalline with formation of extended and parallel chains. Chain orientation about the long chain axis as well as rotation may begin gradually as the temperature increased. For relaxation, chain flexibility is more important for sufficiently long segments. The primary relaxation or motion is the initiation of the micro-Brownian motion of the molecules belongs to the amorphous region. This occurs at glass transition temperature. This is related to the co-operative motion of the fairly large length of molecules. This relaxation is due to the molecular process of conformational changes in a frictionally interacting molecular field in the amorphous region. The secondary absorption is associated with the local twisting motion of the main chain or flexible side chain. The low temperature secondary absorption is associated with side chain motion, frozen amorphous region and might be with the defect region in the crystalline phase. Molecular chains under less restraint in the crystalline texture initiate their thermal motion at a lower temperature or at a shorter time, while molecular chains under greater restraint initiate their thermal motion at a higher temperature or at a longer time.

William, Landel and Ferry equation, commonly known as WLF equation describes the temperature dependence of the relaxation process of polymers. The WLF equation is as follows:

$$\ln \quad\quad \alpha_T = \ln \tau/\tau_1 = [C_1 (T - T_s)]/[C_2 + T - T_s] \quad\quad (4.27)$$

where τ/τ_1 is the ratio of the corresponding relaxation times at two temperatures, i.e. T_1 and T_2. C_1 and C_2 are two empirical constants. T_1 is taken to be nearer to T_g. These constants are mostly same for many polymeric materials. These constants can be related to the free volume fraction, f.

$$f_o = f_1\{1 + \alpha_f (T - T_o)\} \quad\quad (4.28)$$

The thermal co-efficient of $f = \alpha_f = df/dt$. The values of $C_1 = 1/2.3$. f_g and $C_2 = f_g/\alpha_f$, f_g is the fractional free volume at T_g. This can be modified as the well-known 'Doolittle' equation

$$-\ln \tau/\tau_1 = [1/f]/[1/f_g] \quad\quad (4.29)$$

For a simplified WLF equation, the constant value is used like $T_s = T_g + 50$, then $C_1 = 20.4$ and $C_2 = 101.6$; $T_s \cong T_g$, then $C_1 = 40$ and $C_2 = 51.6$. WLF equation may be summarized as:

1. This equation shifts the relaxation spectrum from one temperature state to another temperature.

2. It indicates f, which is a characterization parameter for the shift in relaxation times.

3. The fractional free volume consists of two parameters, i.e. free volume and occupied volume.

4. Both free volume and occupied volume are temperature dependent above T_g.

5. It also indicates another empirical parameter, occupied volume.

The free volume is a parameter for characterization of physical behaviour. The occupied volume is the volume of the crystal lattice and it is practically constant in the glassy region and at glass transition temperature. The main transition is the glass transition. Apart from this main transition, there are minor transitions in the glassy region. These minor transitions are referred as secondary transitions. They are usually labelled from higher to lower temperatures with successive letters of Greek alphabets like α, β, γ, δ, etc. The glass transition temperature is usually denoted as α. There are different motions possible below glass transition temperature. Different polymers and/ or fibres exhibit different molecular motions or relaxations. The notations used are not fixed for any particular type of molecular motions. Depending upon the polymer or fibre and temperature, a particular type of relaxation may be expressed as different notations. At glass transition temperature or below glass transition temperature larger parts of the main chain are frozen. However, all these transitions require a considerable amount of free volume. It is supposed that below glass transition temperature, the available free volume is still enough for the motion of the smaller groups. A schematic diagram of some of the molecular motions is discussed below in Fig. 4.3.

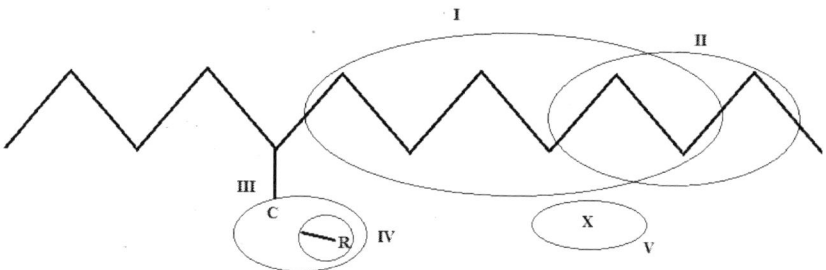

Figure 4.3   Structural representation of molecular motions in glassy state.

I.   This is a motion, within the polymer chain, but locally much more restricted than the motion corresponding to glass transition temperature. This motion marks the transition from brittle behaviour to tough behaviour.

II.   This motion is related to crankshaft motion of the carbon–carbon atoms. Some polymers having a continuous carbon chain exhibit crankshaft motion provided there will be at least 4 carbon atoms.

III. This type is the rotation of a side group about the bond linking it to the main chain. The side group moves as one unit. Its rotation need not be complete and it is more likely to be a transition from one equilibrium position to another.

IV. This type of motion is an internal motion within the side group itself without any interaction with the main chain.

V. This is a motion of a small molecule (*X*) dissolved in the polymer. This can be a motion within a plasticizer molecule or any other additive molecule.

VI. Small molecules dissolved in the polymers, sometimes associated themselves with side groups. If the side groups and the dissolved molecules are both polar, then there will be association between these two groups. This will result in a combined motion at some other temperature instead of motion at 5.

The different motions are dependent upon the structure of the polymer of the fibre. This can be single bond rotation to the motion extending over the whole contour length of the polymeric material. There exist various states of aggregation of molecular chains, which influence the molecular motion. The relaxation is associated with the twisting motion of long side chains, followed by micro-Brownian motion of the side chain and micro-Brownian motion of the main chain. The relaxation below glass transition temperature is also associated with some molecular motions in defects of lamellar crystals and with frozen interlamellar non-crystalline regions. However, this type of molecular motion will be absent in purely amorphous materials. Apart from this, there might be different relaxations present in the amorphous region of a semi-crystalline material and purely amorphous material. The amorphous region in a semi-crystalline polymer is different from that of completely amorphous polymer in that the molecular motions are greatly restricted by the crystalline component.

The internal molecular motions can be analysed by means of dynamic mechanical spectroscopy, nuclear magnetic resonance spectroscopy, neutron scattering, dielectric loss, ultrasonic waves and thermal expansion.

4.9 Properties dependent on amorphous region

Amorphous region is characterized by more random molecular chains with minimum intermolecular forces. So this region is more susceptible to all types of activities or penetration of any additives like moisture, water, steam, chemicals, solvents, dyes. In turn this region strongly affects (a) mechanical response to the imposed force, (b) transport process and (c) diffusion kinetics of different chemical agents. Some of the properties of the material are governed by these penetrations and these are discussed below.

1.   Moisture regain

Water can penetrate inside the amorphous region, and depending upon the sites, the water molecules will be absorbed on the sites of the chain molecules. Penetration of water is basically a diffusion process but as the molecular weight of water is very small, hence the penetration of the water molecules is quick. However, it penetrates inside the amorphous structure of the fibre. The fraction of water molecule absorbed by the fibre is termed as moisture absorption or moisture regain. It depends upon the chemical structure as well as active sites present.

2.   Swelling

Water can penetrate quite readily the amorphous region and since the chain molecules are oriented along the fibre axis, this water will swell the fibre mainly in the lateral direction than in the longitudinal direction. The lengthwise swelling of nylon, cotton and wool is approximately 1.2%, and diameter swelling of these fibres is 5%, 14% and 16%, respectively. On the other hand, the swelling for viscose fibre is 3–5% lengthwise and 26% in diameter. The amount of swelling which occurs depends on the amount of amorphous material, the size of the crystallites and the presence of the polar groups in silk than nylon, so its swelling is much greater.

The total volume of the fibre and water taken together decreases after water absorbed. If a graph will be drawn by plotting density as measured in a non-swelling medium like xylene has been plotted against regain, it will be seen that there is an initial rise of density followed by a fall at higher regains. This is explained by the idea that there is some free volume in the fibre. If the density is measured in water, this increase is not apparent and the graph will be different.

3.   Heat sorption

When fibres absorb water, they evolve heat. It is importance since it has a thermostatic action tending to keep the wearer warm when moisture is absorbed and cool when disordered. The differential heat of absorption of water at zero moisture is almost constant for all fibres between 200 and 300 cal/g, a value roughly same as heat of hydration of OH groups on carboxyl ions.

The definition of two heats of absorption are: (1) differential heat: heat evolved by an infinite mass of material at regain on absorbing unit mass of water and (2) integral heat: heat evolved to dry material in absorbing sufficient water vapour to raise its moisture content from 0 to a particular value.

4. Dyeing

Dyes are basically macromolecules and applied on the fibre from an aqueous medium. Depending upon the size of the dye and structure of the fibre, the dye first absorbed on the surface of the fibre and then diffuses exclusively inside the amorphous regions of the fibre. The dye absorbed by the fibre, and its fastness property is dependent upon the fraction of the amorphous region as well as its orientation.

Further readings

1. J. E. Mark (Ed.), *Physical Properties of Polymer Handbook*, Springer, 2007.

2. L. H. Sperling, *Introduction to Physical Polymer Science* (2nd ed.), Wiley, New York, 1992.

3. J. W. Hearle and R. H. Peters (Eds.), *Fibre Structure*, Butterworth, London, 1963, p. 346.

4. *Encyclopedia of Polymer Science and Engineering*, Wiley, New York, 1986.

5. Z. H. Stachurski, *Fundamentals of Amorphous solids*, Wiley-VCH, 2015.

5.1 Molecular architecture

Polymer is a long chain, mostly linear molecule, with few thousand atoms present in its long chain. A common mean of expressing the length of the chain is the degree of polymerization, and it denotes number of mer or the number of atoms incorporated in the chain. The length of the chain molecules is dependent upon the degree of polymerization and the length of the repeat unit. The length can be approximately 500–1500 Å. The crystallite length ranges from 80 to 150 Å. The structure of any polymer or fibre is generally present in the following way:

1. A number of small molecules joined together to form a long chain molecule, like that of a chain link.

2. The atoms are linked in the molecule by means of covalent bonds.

3. The molecular chains are flexible to assume any conformation.

4. Long chain molecules are arranged together.

5. Some molecular chains lie parallel and sufficiently close together to act as a unit, known as crystallite.

6. Some molecular chains are arranged at random to form amorphous region.

7. Some long molecular chains run through the amorphous region and also two or more crystallites to form a network, which holds the structure together.

The molecular chains, when ordered can be crystalline otherwise, it will be random or amorphous. Based on these facts of polymer molecular structure, morphology deals with physical organization of the macromolecule and it is the overall form of the structure. Morphology deals with structure on the microscopic level. The simplest concept of morphology treats the molecular aggregation as two distinct phases – a nearly crystal phase imbedded in an amorphous entanglement of chains with many chains joining the two phases.

5.2 Two phase structure

The semicrystalline polymers or fibres consist of two phases, i.e. crystalline phase and amorphous phase. Based on the facts of the length of the chain molecules and the dimensions of the crystallites, different models were evolved. There exist a number of models for the description of the superstructure. The co-existence of crystalline and amorphous regions, being treated in the extreme as two separate phases, has provided the basis for various models or theories of fibre structure since 19th century.

Detailed examination of the crystallites, amorphous region and the chain molecule reveals that the chain molecular length exceeds the crystallite length by at least two orders of magnitude. Also the crystallite dimensions are independent of molecular length. Based on these facts different models were evolved. The fringe-micelle model is the earliest one. As per this model, each chain molecule is considered to thread its way through several adjacent crystalline and non-crystalline regions. So the crystalline regions are tied together by segments of molecules, which involve primary valence bonds. Another feature of this model is that the amorphous region is continuous. It can be assumed that the amorphous regions are acting as matrix in which the crystallites are embedded. Both the regions are co-existing in this manner and interact with each other. Each crystallite has its own size and so has its own melting point. This concludes that the melting behaviour of this model is the gradual melting of different crystallites.

The simplest one contains one-dimensional stacks of lateral infinite plate lamellas. The next one is quasi-one-dimensional finite blocks of different shapes. Many workers began considering fibres are fully crystalline materials with varying levels of crystalline imperfections. Molecular dislocations, impurities and chain ends are the principal sources of these imperfections which form a chemical accessibility as well as from a mechanics point of view provide the means of interpreting the macroscopic properties previously associated with amorphous region.

The crystallite size is usually higher for a sample crystallized from solution (more than 100 Å). On the other hand, if it is crystallized from melt, then it would be less than 100 Å. The amorphous region length can be around 20–60 Å, depending upon the crystallinity. This analysis of the crystallites, amorphous region and the chain molecule reveals that the chain molecular length exceeds the crystallite length by at least two orders of magnitude. Also the crystallite dimensions are independent of molecular length. The crystalline chain portion tends to become fully extended, because they are connected covalently to the non-crystalline chain portion, they tend to extend the non-crystalline chain portions. But the non-crystalline portion is more random in space and the associated entropy force tends to pull the crystalline portions apart laterally.

Some chain folding is a topological necessity in every crystalline or lamellar structure. The chains present in the crystalline region changes abruptly into an amorphous region after it come out from the crystallite. Each molecular chain present in the crystallites leaves the crystal and goes into the amorphous region. In the crystalline region, the chains are all parallel and oriented in the same direction. The amorphous region, the molecular chains have a tendency to form random coil. This may form bridges between adjacent crystalline chain molecules. The random coil portion of the molecular chain forms. On the other hand, the loops, which are far more numerous than bridges, are not counted as adjacent re-entry folds.

5.3   Models of fibre structure

The two dimensional models treat the molecular aggregation as two distinct phases – a nearly crystal phase imbedded in an amorphous entanglement of chains with many chains joining the two phases. The model of fibre structure differs in their description of the fine structure of microfibrils and fibrils. The micellar theory for the structure is the earliest one. Here the structure is visualized as that of the polymer chain molecules present as 'micelles'. The crystalline region frequently referred as 'micelle' was taken to be a chemically inaccessible domain. The micelle is thought to be bundles of rigid molecules and contained 40–50 closely packed chain molecules. The micelles are regions of high order spastically distributed in regions of low order. However, this theory was not successful. As molecular dimensions became better established by both chemical and physical methods, it became evident that a polymer molecule could not be accommodated in one discrete crystallite or micelle; this is in turn gave rise to the concept of a fringe micelle.

5.3.1 The fringed-micelle model

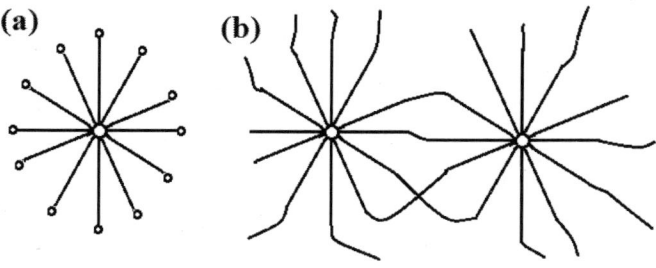

Figure 5.1   Concept of (a) Micelle and (b) Fringe micelle.

Micelle is a spherical formation caused by an amphiphilic substance in a solution. The schematic representation of the micelle is exhibited in Fig. 5.1(a). The lyophilic end of the molecule tends to orient itself towards the outside of the sphere while the lyophobic end tends to orient itself towards the inside of the sphere. Amphiphilic is related to a molecule having a polar water soluble group attached to a non-polar water insoluble hydrocarbon group. Lyophilic and lyophobic are the terminology related to dispersed phase, where lyo means solvent, philic is loving and phobic is hating. Fringe relates to the extra molecular chains coming out from specific objects and attached to a separate object, as shown in Fig. 5.1(b).

The fringe-micelle model (Fig. 5.2(a)) postulates that each chain molecule is considered to thread its way through several adjacent crystalline and non-crystalline regions. So the crystalline regions are tied together by segments of molecules, which involve primary valence bonds. Another feature of this model is that the amorphous region is continuous. The amorphous regions are acting as matrix in which the crystallites are embedded. The amorphous molecular chains are regarded as very long flexible entangled chains, continuously changing their shape due to thermal motion of the units. Both the regions co-exist and interact with each other. Each crystallite has its own size and so has its own melting point. There is no sharp boundary between crystalline and amorphous regions and the postulation of regions of intermediate levels of order. The amorphous region was no longer considered to be a liquid like matrix. This concludes that the melting behaviour of this model is the gradual melting of different crystallites.

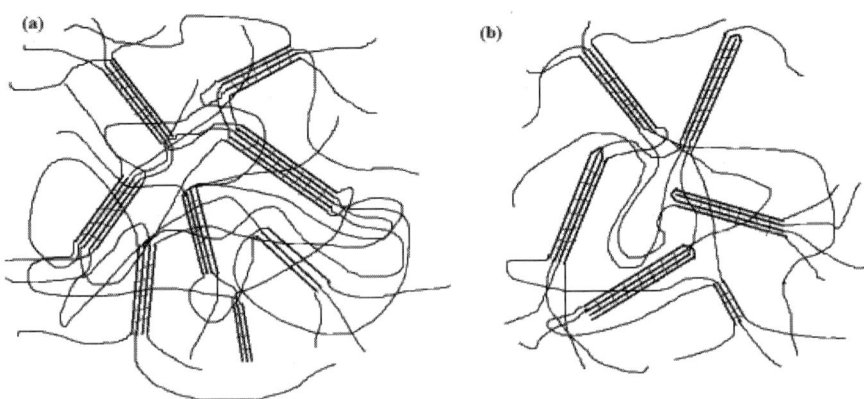

Figure 5.2 (a) Fringe micelle model and (b) Modified fringe micelle model.

With the discovery in 1956 that polymer molecules could fold on themselves, the passage of a polymer molecule from a crystalline region to an amorphous one was no longer a necessary means of accommodating molecular dimensions. The fringed micelle model was modified to fit with better concepts like chain folding, a natural process in the crystallization behaviour of polymers. So a modified fringed micelle model (Fig. 5.2(b)) was introduced in which some of the chain folds back at the edge of the micelle, instead of contributing to the fringe.

5.3.2 The fringe-fibril model

Fringe-micelle model assumed that the polymer chain included segments precisely aligned over distances correspond to the length of the crystallites and more disordered segments belonging to the amorphous region. But chain molecules are very long and can be visualized as passing through several crystalline and amorphous regions. This leads to inadequacy of the fringe-micelle model and to a new concept of fringe-fibril model. Fibrils are basically a threadline structure. Fringe-fibril model is one form of the fringe-micelle model, in which the dimensions along the chain axis is much greater than the two perpendicular directions (Fig. 5.3). Long chin molecules in fibres or oriented polymers are passing through many small crystallites, where molecules packed in a regular order. The size of the crystallites is of the order of 500 × 50 Å. In between crystallites, the molecules pass through non-crystalline regions, where the arrangement is more open and disordered. The length of the fibril depends upon the polymer type and processing conditions. The fringed fibril model is also modified to allow some chain folding. However, the extent of chain folding is smaller than that of the micellar model because of the needle-like characteristics of the fibrils.

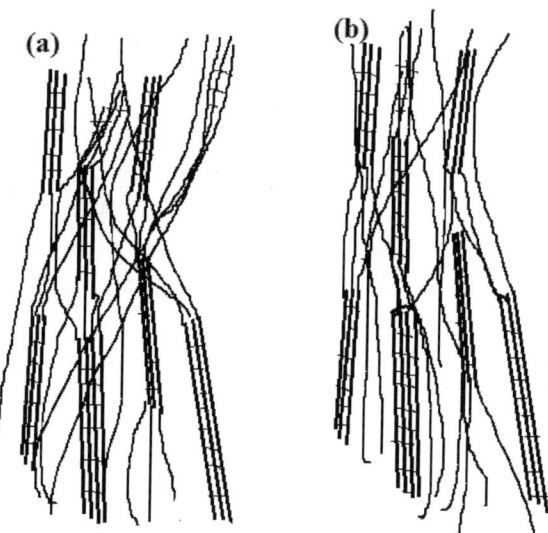

Figure 5.3   (a) Fringe fibrillar model and (b) Modified fringe fibrillar model.

5.4   Three phase structure

As per two-dimensional models, the molecules interchanged the positions from crystalline region to the amorphous region, i.e. from an ordered structure to a random structure. This transformation cannot be abrupt and it can be assumed to be gradual. This means that the fibre structure consists of three components, i.e. crystalline region, liquid like amorphous region and an intermediate component with an interfacial structure. The interfacial region ranges from 5% to 15%. The interfacial region is very diffuse, distorted and many units thick with a structure, which changes gradually between those of the two regions that, it connects. The boundary is neither sharp nor clearly defined. The amorphous region consists of chain units in non-ordered conformations, which connects crystallites. Its properties are very similar to those of pure liquid polymer. The chain units connecting crystallites are not tie-molecules. Only portions of molecules are involved, which are not fully extended. In summary one concludes that a diffuse interfacial zone is associated with crystallite and that an amorphous over layer connects successive one. In a drawn fibre, the molecular chains are highly extended and essentially parallel to each other but chain folding exists to produce a low overall statistical order. In this state, there is a broad distribution of intermolecular bond distances and bond directions with a consequent bond distribution and bond energies. The order in the crystalline regions can be well defined by x-ray diffraction but the amorphous regions cannot be characterized completely as the order in the amorphous region varies from random to highly ordered.

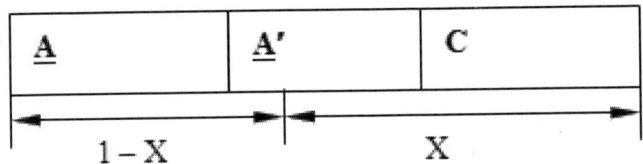

Figure 5.4   Schematic figure of different regions: (A) Amorphous region, (A'), the defect region in the lamella, (C) perfect crystal region and (X) crystallinity.

The highly ordered amorphous region may be the defect region in the lamella as shown in Fig. 5.4. In this figure (A) is the amorphous region segregated from the lamellar phase, (A'), the defect region in the lamella, (C) is the perfect crystal region and X is the crystallinity. The lamellar structure also includes a large defect region. Depending upon its order, the part of the region is treated as crystalline and part as amorphous.

This gives to the concept of tie chains or Taut Tie Molecules (TTM). Taut tie molecules are the bridges between the amorphous layers and the crystal blocks postulate continuous models like (a) Continuous amorphous structure with embedded in crystalline regions or (b) Continuous crystal model with randomly distributed amorphous inclusions. The tie molecules are generally taut and present on the boundary of the microfibrils. All the intrafibrillar tie molecules are taut and arranged on the other boundary of the microfibril. The plastic deformation process increases substantially the fraction of interfibrillar tie molecules. The tie molecules are considered as an important aspect of fibre structure and it influences the deformation properties of the fibre.

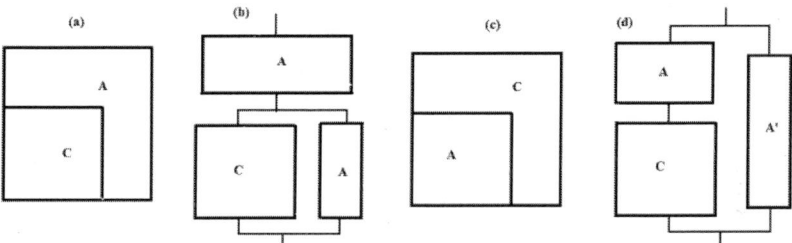

Figure 5.5   Schematic figure of unit cube models: (C) Crystalline region, (A) amorphous region, (A′), TTM.

The arrangement of the taut tie molecules can be understood and explained by means of unit cube models. The schematic diagram is shown in Fig. 5.5. The continuous amorphous structure with embedded in crystalline regions is shown in Fig. 5.5(a). The respective model is represented in Fig. 5.5(b). The continuous crystal model with randomly distributed amorphous inclusions is shown in Fig. 5.5(c) and the respective model is shown in Fig. 5.5(d).

5.5   Crystalline relaxations

The ideal crystalline structure is a single crystal with no defects and having maximum stability. However, the fibres consist of crystalline region and amorphous region. The crystalline region consists of crystallites having different size and shape, crystal defects, interlamellar regions and a varied degree of crystal stabilities. Owing to this, crystalline regions show relaxation in the temperature region close to the melting range. The relaxation mechanism can be related to interlamellar slips or the movement of the intermosaic block boundaries. In some polymers, there will be more number of crystalline relaxations. These are termed as $\alpha_I, \alpha_{II}, \alpha_{III}$ in ascending order of temperature.

The interlamellar slip may relax at a temperature nearer to the glass transition temperature. This induces a conformational change of the molecular chains belonging to the interlamellar region. This gives a view that there is an intermediate region in between the crystalline and amorphous region. Sometimes the crystalline region is attributable to the oriented molecular chains. These molecular chains are parallel to the molecular chains present in the crystals. However, the molecular chains have some distorted conformations and it restricts the molecular chains to be present with the crystallites and so these are not fitted with the crystallites.

Some relaxation phenomenon arises from directly the crystalline region. This relaxation is dependent upon the long period of the polymeric structure. There can be various motions of molecular chains like rotational motion or translational motion along the molecular axis. Crystalline absorption of single crystal is affected by the morphological factors of the crystal and its perfection. This is because of the modes of the torsional motion of the molecular segments constituting different lamellar thickness. The single crystal thickens by heat treatment above its crystallization temperature. The spherulites are composed of piled thin plate like crystals, i.e. lamellar with tie molecules between neighbouring lamellae in the spherulites. The lamellar phase of spherulites include defect region also.

5.6 Crystallinity

The amount of crystallites or the crystalline regions can be expressed by the term crystallinity. Crystallinity indicates the presence of three-dimensional orders on the level of atomic dimensions. The polymer showing or having crystallinity is referred as crystalline polymer. The quantitative expression for crystallinity is the degree of crystallinity and it indicates the fractional amount of crystallinity in the polymeric sample. The degree of crystallinity may be expressed either as the mass fraction crystallinity, w_c or x_c or as the volume fraction crystallinity ϕ_c. These two fractions are related by means of Eq. (5.1).

$$w_c = \varphi_c \cdot \rho_c / \rho \tag{5.1}$$

where ρ_c and ρ are the densities of the crystalline sample and the entire sample, respectively. The measurement of the crystallinity made with the assumptions that the sample can be subdivided into a crystalline phase and an amorphous phase as per two phase model theory. Both phases are assumed to have properties identical with those of their ideal states, with no influence of interfaces. The degree of crystallinity can be determined by several experimental techniques. The most commonly used techniques are: (1) x-ray diffraction, (2) calorimetry, (3) density measurements, (4) infrared spectroscopy. The methods of measurement of crystallinity by these methods are given below. However, the detail methods of measurement are discussed in detail in the respective chapters.

5.6.1 X-ray diffraction

The degree of crystallinity ($X_{\text{X-RAY}}$) can be calculated as per Eq. (5.2):

$$X_{\text{X-RAY}} = \frac{\sum_{S_1}^{S_2} S^2 I_c(S)}{\sum_{S_1}^{S_2} S^2 I(S)} \tag{5.2}$$

where $s = 2/\lambda.\,(\sin\theta), \lambda$, the wavelength, θ, Bragg's angle. This equation requires a mathematical computation of the intensity obtained by the sample and its crystalline component. However, the detail of crystallinity measurement is given in Chapter 9.

5.6.2 Density measurements

The degree of crystallinity (X_D) can be calculated by density methods as per Eq. (5.3)

$$X_D = \frac{\rho_c(\rho - \rho_a)}{\rho(\rho_c - \rho_a)} \tag{5.3}$$

where ρ, ρ_c and ρ_a are the densities of the sample, of the completed crystalline sample and of the completely amorphous sample, respectively.

5.6.3 Calorimetry

The degree of crystallinity (X_{DSC}) can be calculated by means of the calorimetric methods with Eq. (5.4)

$$X_{\text{DSC}} = \Delta H_{\exp} / \Delta H^\circ \tag{5.4}$$

where ΔH° is the heat of fusion of fully crystalline sample and $\Delta H_{\exp}$ is the heat of fusion of the sample. The value of ΔH° may be obtained by extrapolating $\Delta H_{\exp}$ to the density of the completely crystalline sample. The value of heat of fusion is also related with density as per the relation: $\Delta H = x - y\cdot v$, where v is the specific volume at 20°C in cm³/g

For PET, $x = 1411$ and $y = 1886$ and for PBT, $x = 1296$ and $y = 1628$.

5.6.4 Infrared spectroscopy

The degree of crystallinity (X_{IR}) can be calculated by density methods by means with Eq. (5.5)

$$X_{IR} = \frac{1}{a_c \rho l} \log_{10}(I_0 / I) \qquad (5.5)$$

where I_0 and I are, respectively, the incident and the transmitted intensities at the frequency of the absorbed band due to the crystalline portion, a_c is the absorptivity of the crystalline material, ρ densities of the sample and l is the thickness of the sample. However, the detail of crystallinity measurement is given in Chapter 11.

Imperfections in crystals are not easily distinguished from the amorphous phase. The various techniques may be affected to different extents by imperfections and interfacial effects. Hence, there will be some degree of disagreement among the results of quantitative measurements of crystallinity by different methods.

5.7   Fibre structure and properties

The organization of the polymer chains in the three-dimensional space determines to a large extent the chemical, physical and mechanical properties of the fibre. Characteristic fibre properties are achieved through the development of an intermolecular chain organization that can be described as highly oriented and semicrystalline. Although the crystalline or highly ordered regions of a fibre have been the targets of structural investigation, the disordered or amorphous regions may indeed play a dominant role. The amorphous region is presumed to consist of regions with a low level of lateral order. The concept of a lateral order distribution, first proposed for cellulosic fibre, may be the best means of describing the three-dimensional organization of polymer chains in semicrystalline or polycrystalline materials like fibres.

At the present time, fibre structure is viewed as somewhere intermediate between the fringed polyphase model and the fully crystalline folded model. The crystalline regions certainly do contain polymer chain folds and imperfections in these regions may be taken to account for many fibre properties. On the other hand, the accessibility of fibres to large and bulky molecules, such as polymers and dyes, does suggest the existence of fairly large domains with at least localized regions of low density. Nevertheless, the existence of polymer chain folds in most synthetic fibres is undisputed and IR spectroscopy evidence of molecular folds in PET and PA has been studied.

Table 5.1 Properties of fibre at higher crystalline structure.

Properties modified at higher crystallinity	
Increase	Decrease
Tensile strength	Extensibility

Initial modulus	Flexibility
Hardness	Toughness
Dimensional stability	Dye absorption
Density	Moisture absorption
	Chemical reactivity

Many properties of fibres are dependent on the degree of crystallinity (Table 5.1). Relationships are complex and must therefore be qualitative. The size and orientation of the crystallites modify the relationships. Most native cellulosic fibres, with crystallinity of the order of 80% have higher dynamic modulus than regenerated cellulosic fibres with crystallinity in the order of 40–60%. The modulus increases with increase in orientation.

5.8   Liquid crystal polymer (LCP)

Liquid crystalline polymers are a special class of polymers where it exhibits crystalline characteristics even at the liquid stage. The arrangement of the molecules lies between regular order of the crystalline structure and the disorder in amorphous or liquid structure. When the material melted or converted into solution with restricted solvent concentration, it exhibits molecular order. It retained the orderly arrangement of the molecules, i.e. parallelization. Based on the medium, where the material shown liquid crystalline characteristics, there are two types of liquid crystalline polymers, i.e. (a) thermotropic liquid crystalline polymers and (b) lycotropic liquid crystalline polymers. Thermotropic liquid crystalline polymers are known as thermotropic because the phase transitions are brought about by the variation of the temperature alone. These materials show liquid crystalline characteristics above melting point and when the temperature is further increased, the material exhibits a normal isotropic liquid with no alignment. This temperature is referred as isotropization or clearing temperature. The lycotropic material exhibits liquid crystalline characteristics in a particular solvent at particular or fixed range of concentration. Here the transition depends on the concentration of the solvent used. The structural requirement for a polymer to exhibit liquid crystalline characteristics is as follows:

1.   The molecules or a part of the molecules should be rigid and rod shaped.

2.   The molecules should be elongated.

3.   The length to diameter ratio, i.e. aspect ratio should be high.

4.   There should be aromatic rings in the system.

5. The rings should be connected with para position.

6. The interconnected links should be rigid.

The rigid rods, which are responsible for the liquid crystalline materials, are called mesogens. The mesophase of the rigid chain liquid crystalline polymers is characterized by parallel arrangement of the molecular chains. The molecular units consist of rigid units and so the chains form near parallel configuration in the molten mesophase. The rigid chain thermotropic polymers can exhibit sharp x-ray diffractogram, associated with crystallites with three-dimensional orders. Such materials show two transitions, i.e. T_g, glass transition temperature and T_m, melting temperature. T_g is associated with the transformation from a very rigid solid to a more flexible solid. T_m is associated with a change to a birefrigent mesophase melt of very low viscosity. The rigid state in the temperature in between T_g and T_m can be attributed to the crystallites linking together to the rigid chain. This is similar to that of the crystalline structure in conventional crystalline polymers. The chain stiffness is the main important property that will determine whether the polymer melt will exist as a nematic mesophase rather than as a conventional isotropic melt. If the melt is nematic, then there will be a low entropy change since there is no change in overall configuration on crystallization.

$$T_m = \Delta H_f / \Delta S_f \tag{5.6}$$

The relationship between melting temperature heat of fusion and heat of entropy is shown in Eq. (5.6). For a given ΔS_f, T_m will depend on ΔH_f, i.e. on the chain regularity and crystal packing. If ΔS_f is low, then ΔH_f must be low, otherwise T_m will be too high. Low ΔS_f results from the nematic state of the melt that in turn is a consequence of the chain stiffness. Low ΔH_f is a reflection of the chain irregularities preventing good crystal cohesion and is an essential feature of the design of the liquid crystal molecules. The thermotropic liquid crystal polymer exhibits the following specialties.

1. The polymer exhibits unique order in the fluid state, controlled either by electric, magnetic, stress, fields or temperature.

2. Melt temperature of this type of polymer require higher temperature than its conventional polymer.

3. The structure and properties can be controlled as cooling of the oriented liquid crystalline structure preserves the structure in the solid state.

4. The inherent orientation created by the force results in high strength as well as properties.

5. The inherent orientation coupled with high strength makes that specific polymer a useful material with unusual performance properties.

Lycotropic polymer also exhibits similar behaviour under particular solvents.

A simple liquid crystalline material is poly p-phenylene as shown in Fig. 5.6.

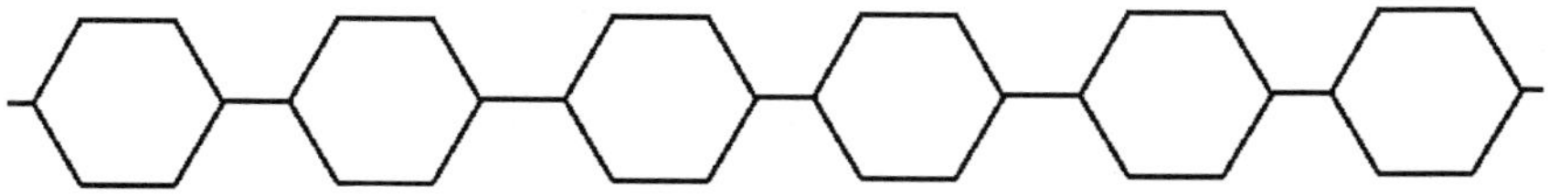

Figure 5.6   Structure of poly p-phenylene (PPP).

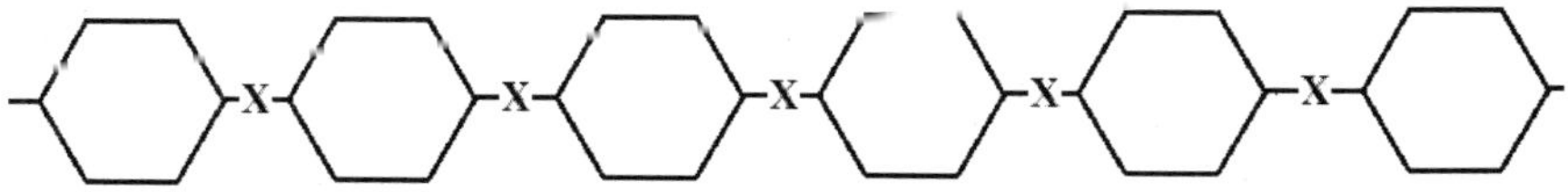

Figure 5.7   Structure based on poly p-phenylene (PPP).

There will be different structures based on poly p-phenylene (PPP). This is shown in Fig. 5.7.

These materials have special application in plastics and fibre. Kevlar and carbon fibre are two fibres, which are produced by utilizing liquid crystalline characteristics nature of these two polymers. In melting stage or in solution stage, the polymer molecules are locally well aligned and closely packed but not as well ordered as in a crystal. The rigidity of the chain practically excludes chain folding. In a flow field with longitudinal and/or transverse gradient the large liquid crystals are easily oriented in flow direction. The lower density of lateral packing makes the longitudinal slip of chain molecules is much easier than crystals with densest possible packing in spite of the fact that the thickness of the liquid crystals may be very much larger than that of the folded chain lamellae. The flow with transverse gradient results in perfect orientation.

Further readings

1. L. H. Sperling, *Introduction to Physical Polymer Science* (2nd ed.),Wiley, New York, 1992.

2. J. E. Mark, *Physical Properties of Polymer Handbook*, Springer.

3. Robert J. Young and Peter A. Lovell, *Introduction to Polymers*, CRC Press, 2011.

4. J. W. Hearle and R. H. Peters (Eds.), *Fibre Structure*, Butterworth, London, 1963, p. 346.

5. W. E. Morton and J. W. S. Hearle, *Physical Properties of Textile Fibres*.

6. Claudio De Rosa and Finizia Auriemma, *Crystals and Crystallinity in Polymers: Diffraction Analysis of Ordered and Disordered Crystals*, Wiley, 2013.

7. *Encyclopedia of Polymer Science and Engineering*, Wiley, New York, 1986.

8. L. Mandelkern, *Crystallization of Polymers*, Mc-Graw Hill, New York, 1964.

9. M. Lewin and E. M. Pearce, *Handbook of Fibre Science & Technology*, Marcel Dekker, New York, 1985.

10. S. Eichhorn, J. W. S. Hearle, M. Jaffe, and T. Kikutani (Eds.), *Handbook of Textile Fibre Structure*, Woodhead Publishing Series in Textiles, 2009.

11. I. M. Ward (Ed.), *Structure and Properties of Oriented Polymers*, Springer, Netherlands, 1997.

12. H. Tadokoro, *Structure of Crystalline Polymers*, Wiley, New York, 1979.

6
Crystallization

6.1 Introduction

All properties in a polymer and/or in a fibre are controlled by their molecular morphology that is the overall structure. The structure of polymer like crystal, crystalline and other ordered and disordered is created by the crystallization mechanisms. This means information about the crystallization mechanisms can be obtained by means of crystallization kinetic studies. So crystallization studies in polymers and/or fibres are important for different properties as well as for different end products. These studies are concerned with the way in which the structure is laid down. The mechanism is mostly concerned with the rearrangement of the molecules to form the more ordered arrays, characteristic of the solid state. Morphology is concerned with the nature of the final state after the crystallization process is complete.

6.2 Crystallization process

Crystallization of polymer is a process associated with partial alignment of the molecular chains which form ordered regions. With this objective, crystallization process is defined as any process, where the structure of the material in terms of crystallinity will be enhanced. For example, the molten polymer or polymer melt does not contain any ordered structure. If the melt is slowly cooled, the temperature will gradually fall because of the temperature gradient. Cooling below the melting point will produce either supercooling of the liquid or solidification of the polymer. Before solidification will be completed, if the material can be hold for certain time at a constant and fixed temperature, then there will crystallization as a result of mobility of the molecular chains and its rearrangement. Generally a definite degree of supercooling must be achieved for crystallization. The crystallization behaviour at the constant and fixed temperature will be almost similar for all polymers and a schematic diagram of the crystallization curve is shown in Fig. 6.1.

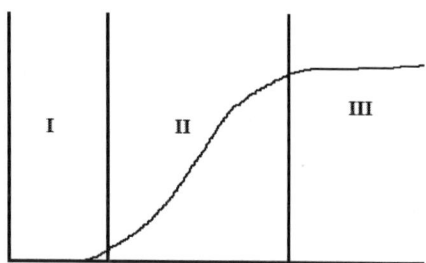

Figure 6.1   Schematic diagram of a crystallization process: (I) induction period (II) primary crystallization and (III) secondary crystallization.

The crystallization behaviour (Fig. 6.1) can be divided conveniently into three portions. The initial portion is (I) induction period followed by (II) primary crystallization and (III) secondary crystallization. The induction period is the time required for the formation of the nucleus for crystallization. The crystallization process of polymers can be of two types. The first one is primary crystallization and the second one is secondary crystallization. Primary crystallization is the first stage of crystallization, where almost all the chain molecules impinge to form the crystallites. Two kinetic steps, i.e. (a) nucleation and (b) growth control the primary crystallization process. These processes are involved in the formation of crystallites or crystalline micelles in polymers. The first stage is the formation of small particles and is generally known as nucleation. The second stage is the growth of the nuclei. During primary crystallization, the nuclei grow until the maximum amount of material is incorporated into the crystalline region. The crystallization can be considered to be complete. These two processes form the most important aspects of crystallization. The secondary crystallization process is a time-dependent process, occurring after primary crystallization. It is either followed by primary crystallization or it can occur for a material already crystallized and preceded at a slower rate.

6.2.1 Nucleation

The crystallization process can be initiated with nucleation from a point, generally termed as 'nucleus'. This stage is termed as nucleation. Nucleation is the first step in the formation of either a new thermodynamic phase or a new structure in a self-assembly or self-organization. It is typically defined as the process that determines the waiting time before new phase of the self-organized structure appears. There are two types of nucleation, i.e. homogeneous nucleation and heterogeneous nucleation.

Homogeneous nucleation involves spontaneous aggregation of polymer chains below melting point. In this crystallization process, the nucleus consists of the same element with same chemical structure. The nucleus may form spontaneously, under suitable conditions, and is assumed to appear in a molten polymer below its melting point at random and instantaneously. It consists of a small region where the local arrangements of polymer chain segments temporarily created. This ordered region of the nucleus is the result of the random segmental relaxation motions and resembles like that of the structure of the crystallites. This region has a critical size. Below the size, the region decomposes back to randomness, but above the size, the region becomes stabilized by crystal growth. They may also appear progressively throughout the course of the process. In general, when the polymer melt is cooled down, the relaxation rate of segments is slowed down. So the concentration of the orderly regions increases and this causes the rate of nucleation to increase strongly with temperature. Homogeneous nucleation is more complicated because (a) the nucleus may be formed at random and at different times and (b) the nucleation may be formed in a system, where the formed nucleus is already growing.

The heterogeneous nucleation process is a simple process where the nuclei are formed by another external element. Heterogeneous nucleation arises from the addition of small seed crystals or small quantities of adventitious impurities either randomly distributed throughout the bulk or localized on a surface. This additive is commonly known as 'nucleating agent'. The nucleation period may be shortened or modified by addition of nucleating agents. Controlling addition of the nucleating agent can control the nucleation. A critical nucleus size must be attained before primary crystallization can begin. The stronger intermolecular bonding forces, the smaller will be the critical nucleus size. In oriented system, the nucleation period may be absent, since the nuclei are formed by the orienting process. Crystallization of a semicrystalline polymer of fibre will not show any induction period due to presence of crystallites before crystallization. These crystallites will act as the centre for nucleation.

The number of nuclei appear during course of crystallization is of primary importance because it determines the final size of the crystallites formed. If the number of nuclei can be controlled, the size of the crystals ultimately formed can be regulated. High nucleation rate can lead to more nuclei for the formation of crystallites and it resulted in smaller crystallites. On the other hand, larger crystallites can be formed at high temperature of crystallization with controlled nuclei formation and slow growth.

6.2.2 Growth

The growth of the nucleus can be assumed to take place in any dimensions like one-dimensional, two-dimensional or three-dimensional. In one-dimensional, it will give rod-like, two-dimensional disc-like and three-dimensional, sphere-like material. In each case, these bodies grow out from primary nucleus produced by chance fluctuations of local chain arrangements and/or aided by other heterogeneous compounds. The linear dimensions of the growing bodies increase with time. The converted phase is generally assumed to be crystalline with higher density. The relationship between the crystallinity of the system, time of crystallization and temperature provides information on the nature of growth process. The growth process can give rise to a variety of morphological units like single crystals, spherulites, fringed structure, etc.

6.3 Polymer crystallization

Polymer crystallization is a nucleation controlled process. The growth will be dependent upon the nucleation rate. In case of polymer crystallization, nucleation and growth usually occur simultaneously. Owing to this, different type of structure can be obtained for the same polymer during crystallization at different conditions because of formation of different entities of nuclei formation. A slow rate of cooling reduces spontaneous nucleation. Hence the material is deposited on a few nuclei and this leads to relatively large crystals. On the other hand, rapid cooling creates more number of nucleus and results in more number of small crystals. With some materials nuclei from almost immediately super saturation are achieved. Others can be cooled without the deposition of an appreciable number of nuclei, so that crystallization does not take place and the liquid gradually thickens to for a very viscose mass, referred as glass, as the temperature is reduced.

The increase in crystallinity occurring after primary crystallization by means of time or temperature is regarded as a recrystallization process. Basically, this is a growth of the crystallites and the existing crystallites act as nucleating agents. During this process, smaller crystallites melt and the larger crystallites grow. This may reduce the number of crystallites present after primary crystallization. Measurements of crystallinity by some methods do not regard small crystallites, i.e. crystals with smaller dimensions as crystallinity. A small size nucleus with stronger intermolecular bonding may be a stable crystallite. This can give assumption that polymers or fibres with stronger intermolecular bonding forces are often presumed to be less crystalline than polymers or fibres having weaker intermolecular bonding forces.

6.4 Theory of crystallization kinetics

6.4.1 Primary crystallization

A simple model of crystallization kinetics was first derived by Avrami to study isothermal crystallization. The model assumes that the growth of rod-like, disc-like or spherical bodies proceeds from primary centres, which are randomly distributed in space. Each individual body grows at a rate proportional to the size of its growing surface. So the proportionality of the rate of crystal formation to the growing surface of the body implies mathematically that the growing surface advances linearly with time. It is assumed that the radius of a disc-like or the length of a rod-like body grows linearly with time. Ultimately the growth units impinge, ceases locally and the mass is converted into a new phase. The degree of crystallinity of the whole specimen at this state will be X. The overall degree of crystallinity X_t of the sample at time t will then be:

$$X_t = X_\infty (1 - p) \qquad (6.1)$$

where X_∞ is the maximum crystallinity and $(1 - p)$ is the fraction occupied by the bodies. The kinetic of growth as measured by means of Avrami's equation is

$$X = X_\infty \{1 - \exp(-z \cdot t^n)\} \qquad (6.2)$$

where z is the rate constant and n is the Avrami's exponent. It is $1 < n < 4$. It can be measured by the graphical equation of $\ln[(X_\infty - X)/X]$ versus $\ln t$. The value of n determines the types of nucleation and growth like whether it is rod-like, disc-like or spherulitic growth from instantaneous or sporadic nuclei. The value given for the growth is 1, 2, 3, respectively, instantaneous nuclei 0 and a sporadic nucleus is 1. Crystallizing polymers do exhibit characteristic features mentioned in Avrami's equation. Experimentally, it was observed that under different conditions of crystallization (like temperature), there is a real change in the mechanism, which can be quantitatively represented by the change in the Avrami's exponent from 2 to 4.

6.4.2 Secondary crystallization

The bulk crystallization process does not always terminate in the way predicted by the Avrami's equation. With many polymers, a secondary process follows the Avrami's or primary process of crystallization (Fig. 6.1). There will be an additional crystallization process. A slower, secondary process, approximately linear in log time, follows the primary crystallization and it is referred as secondary crystallization. During this process, the enhancement of the crystallinity is slower. Secondary crystallization may be due to the impurities rejected during primary crystallite growth or by melting of marginally stable crystallites. Nucleus, when formed in the vicinity of a crystallite surface is termed as secondary nuclei. The distinction between two phases is often not very apparent but the first part is fitted to Avrami's relation. The crystallinity-time relationship in secondary crystallization can be expressed as:

$$X_t = X_0 + D\log(t - t_0) \qquad (6.3)$$

where X_t is the crystallinity at time t, X_0 the crystallinity at time t_0 and t_0 is the time origin of the secondary stage.

6.5 Crystallization rate

The crystallization rate can be estimated either by solvent crystallization process or by melt crystallization process. The rate of growth of a crystal in a solution is dependent on the temperature and concentration of the liquid at the crystal face. This crystallization process involves mass transfer and also heat transfer. The melt crystallization process involves heat transfer from the crystal surface to the medium. Since the resistance to heat and mass transfer lies predominantly in the laminar sub-layer close to the surface of the crystal, the rate of growth of the crystal is improved by increasing the relative velocity between the solid and the liquid. The rate of crystallization is a function of the degree of saturation. For crystallization from a melt, the process is dependent on the rate of transfer of heat between the crystal face and the surroundings, as shown in Fig. 6.2. If T_m is the melting point and T_l is the temperature of the liquid, the apparent supercooling is $T_m - T_l$. Because of the conversion of latent heat into sensible heat as the material solidifies, the temperature at the interface rises to T_i and the true supercooling is then only $T_m - T_i$.

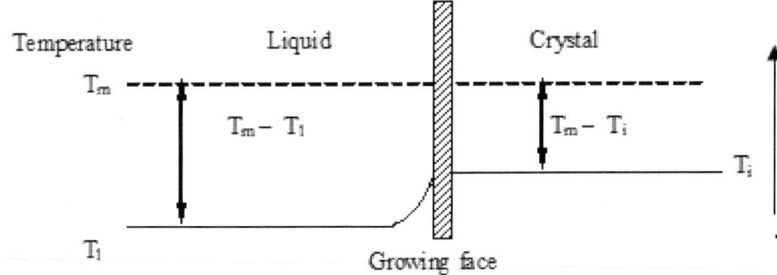

Figure 6.2   Temperature distribution during crystallization.

The temperature at the interface must be greater than that of the bulk of the liquid otherwise there would be no transfer of heat. Higher the heat transfer coefficient result in smaller temperature difference $(T_i - T_l)$ and therefore a high rate of agitation will cause T_i to approach T_l. Although there does not appear to be any regular relationship between the rate of growth of a single face of a crystal and the supercooling, there is a general tendency for a rise in the growth rate as the temperature difference is increased.

For crystallization from a solution, very much less supercooling is possible. The supercooling can be about 2–3° as compared with 50–100° for melts as because the process involves mass transfer rather than heat transfer. In this case, two processes of nucleation and growth cannot be conveniently separated. The solvent crystallization process is shown schematically in Fig. 6.3. C_B is the concentration of the saturated solution and C_A the concentration in the bulk of the liquid. The concentration falls from C_A to C_B through the liquid and the concentration difference $C_B - C_S$ is required in order to overcome the resistance at the interface. Thus whereas $C_A - C_S$ is the apparent super saturation, the concentration difference which is responsible for mass transfer is only $C_A - C_B$. In this case, the heat effects at the face are generally insignificant.

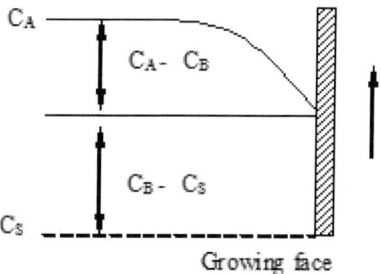

Figure 6.3   Concentration distribution during solvent crystallization.

Although there is no relation for the instantaneous rate of growth of a crystal face, one can obtain a statistical relation over a period of time. The resistance of the laminar layers near the face controls the diffusion of material to the crystal face almost entirely. On the basis, we can say,

$$dM / dt = D \cdot A / \delta_b (C_A - C_B) \qquad (6.4)$$

where A is the area of the crystal surface; D, diffusivity of the solute in the solution; δ_b, thickness of the laminar layer (it decreases as the degree of agitation is increase); and M, mass of material deposited in time t. Again the mass transfer at the interface was directly proportional to the super saturation at the interface,

$$dM / dt = K(C_B - C_S) \qquad (6.5)$$

where K is the rate constant of the surface possess.

Comparing Eqs. (6.4) and (6.5) to eliminate C_B, in both the equations, the value of dM/dt is shown in Eq. (6.6) and the linear rate of growth is shown in Eq. (6.7)

$$dM / dt = \{A(C_A - C_S)\} / \{1 / K + \delta_b / D\} \qquad (6.6)$$

$$\delta l / dt = \{1 / \rho_s\} \cdot (C_A - C_S) / \{1 / K + \delta_b / D\} \qquad (6.7)$$

where ρ_s is the density of the material.

If $1/K$ and δ_b/D are both presumed to be constant, Eq. (6.7) can be modified and shown in Eq. (6.8)

$$\delta l / dt = K_1 \cdot \Delta C \qquad (6.8)$$

where $C_A - C_S = \Delta C$, K_1 is a constant and $\delta l/dt$ is independent of actual size of crystal.

During growth, the shape of the crystal does not alter and therefore the large faces must grow less rapidly than the small ones. This is further evidence of a resistance to transfer at the crystal faces, because, if all the resistance were in the liquid film, each face would grow at the same rate. It is the variation of K from face to face, which allows the crystal to maintain its shape.

The ratio of the volume, to the cube of the linear dimension, must remain constant, as the shape does not alter.

i.e.
$$V = K_2 \cdot l^3 \qquad (6.9)$$

or
$$\delta V / \delta l = 3K_2 \cdot l^2 \qquad (6.10)$$

Thus using the approximate relationship given by Eq. (6.8), the rate of increase of volume is given by

$$\delta V / dt = 3 K_1 \cdot K_2 \cdot l^2 \cdot \Delta C \qquad (6.11)$$

$$\delta V / dt \propto l^2 \Delta C \qquad (6.12)$$

From Eq. (6.12), the rate of growth can be calculated approximately. In practice there might be considerable variations between the experimental value and this equation. Because the linear rate of growth of a crystal is independent of its size, the change in the size distribution of a set of seed crystals can be calculated on the assumption that the size range of these seeds is sufficiently small for the solubility of each crystal to be approximately the same. Thus the seed crystals have dimensions in a particular direction of l_1, l_2, l_3, etc. their volumes will be $K_2 \cdot l_1^3$, $K_2 \cdot l_2^3$, $K_2 \cdot l_3^3$, etc. If during the period of growth of these crystals, this dimension has increased by an amount of Δl, the volumes of the crystals in the product will be $K_2 (l_1 + \Delta l)^3$, $K_2 (l_2 + \Delta l)^3$, $K_2 (l_3 + \Delta l)^3$, etc. Then the total mass M of the solute crystallized can be estimated as per Eq. (6.13)

$$M = \rho_c \cdot K_2 \{\Sigma(l_1 + \Delta l)^3 - \Sigma l_1^3\} \qquad (6.13)$$

This expression enables the change in the size distribution of the crystals to be calculated after a known amount of solute has been removed from solution.

6.6 Types of crystallization

The crystallization of polymers can be of three different types. Those are:

1. Crystallization of long chain under quiescent conditions, i.e. (a) melt crystallization and (b) solvent-induced crystallization.

2. Crystallization of long chain induced by orientation, i.e. strain-induced crystallization.

3. Crystallization concurrent with chain growth

6.6.1 Melt crystallization

Melt crystallization is related to the crystallization from melts. This is a condensed system. Lowering its temperature to below melting point solidifies the molten polymer. During this melt crystallization process, the configuration of the chain remains unaltered and the chain merely 'freezing in' on crystallization. So it is more concerned to the mechanism by which the molecules in the supercooled liquid rearrange themselves to form more ordered structures. At any time or temperature during cooling the growth of crystals at some extent is generally stopped because of restrictions imposed on chain mobility of the molecules. The structure will have both crystalline and non-crystalline regions in a manner like fringed micelle model.

During the crystallization process, the chain molecules reorganize themselves and results in chain folding. Under suitable conditions, crystallization will lead to higher crystallite thickness. Samples crystallized at low supercooling will result in thickening of the crystals at the crystallization temperature. At low supercoolings, the chain molecules can move over long distances and can organize themselves into regular textures in highly specific ways. The structure thus formed is more perfect and regular with chain folding. The folding and fold structure are regular with systematic adjacent re-entry. However, samples crystallized by rapid supercooling are hardly the most suitable ones on which the general rules for crystallization applies. The fact is that the chains will not have a chance to arrange themselves in their most preferred fashion. The chain molecules will not be able to form and pass into different morphological regimes at high undercoolings.

A lower temperature crystallize morphology is described as folded chain lamellar structure. The crystal of this morphology described as the polymer chain passes through a lamella only to emerge at the other side of the fold and return back similarly. The chain might travel back and forth resulting folding. The lamellae have been referred to as thermally induced crystals.

The unoriented crystalline polymer exhibits well developed spherulitic structure. In a quenched sample, there may not be any fully developed spherulites. But there will be stacks of densely packed parallel lamella. This sometimes referred as micrspherulites. The stacks are randomly oriented because of random orientation. If the number of nuclei per point volume is smaller, as in case of isothermal crystallization, close to the melting point or in crystallization during slow cooling, the stalks continue to grow and develop into spherulites by non-crystallographic branching of lamellae.

6.6.2 Solvent-induced crystallization

Solvent-induced crystallization is related to crystallization in presence of solvents. This is a diluted system. The crystallization generally proceeds in a heterogeneous medium, i.e. in presence of additives like solvents or any nucleating agents. The addition of these solvents or nucleating agents creates heterogeneous nucleation. The major difference between solvent-induced crystallization and thermal crystallization is that crystallization takes place in presence of another molecular species, i.e. solvent and therefore in a swollen state of the polymer. The presence of certain interactive solvents or liquids can induce crystallization even at room temperature. The interaction of the liquids with the polymer depends upon the characteristics of the solvent and their interacting capabilities. Some highly interactive liquids also penetrate inside the crystalline structure, in addition to its interaction with the amorphous structure. In general, the solvent enters in the polymer chains, weakens polymer–polymer interaction, replaces with polymer–solvent interaction, induces extensive segmental motion and lowers the effective glass transition temperature of the material. The polymer chain will rearrange them into a lower free energy state. This induces crystallization in the swollen state. The resulting crystal entities are usually isolated and can be separated from the surrounding medium. So it is usually expected that the chains are perpendicular or at a large angle to the basal plane and also chain folding is a straightforward necessity. The nuclei are fixed as the amounts of the additives are fixed. Presence of the solvent induces high nucleation rate, and it results in high concentration of small crystallites. The growth of individual crystallization centres like spherulites or crystallites is a steady state surface nucleation process. Increase in crystallinity by means of solvent-induced crystallization is practically rate dependent. The process is controlled by the diffusion of the polymer. The extent of the crystallization is proportional to the penetration distance of the liquid.

When crystallization was performed in dilute solutions, the molecules do not penetrate each other and thus crystallize independently. This leads to crystallization of the individual molecular chains. It assumes a statistically coiled chain conformation in solution. This is generally referred as 'stacked-sheet' model. Parts of the molecule stick on the growth surface and are incorporated into the growing crystal. If one assumes that only one molecule crystallizes at any instant, one would expect a collapsed Gaussian conformation resulting from a compressed and crystallized molecule. This should be of an elliptical shape. The solution-crystallized sample creates more imperfect crystallites as well as smaller crystallites.

Spherulitic structures are characteristic of the morphology induced by solvent treatments in unoriented amorphous polymers. So the solvent-induced crystallization leads to formation of spherulites. The crystallization rate can be assessed by the change in spherulites radius. The rate of change of the radius of the growing spherulites (G) can be ascertained by means of the Eq. (6.14)

$$G = G_o \cdot \exp(-\Delta E_D / RT) \cdot \exp(-\Delta F^* / RT) \qquad (6.14)$$

where G_o is a constant and equals to $J_o \cdot b_o$; J_o, frequency factor; b_o, jump length for the crystallizing unit in the supercooled melt; ΔE_D, activation energy for transport of the crystallizing unit across the crystal/melt interface; ΔF^*, free energy barrier for the formation of a stable crystal/melt interface. If J is the jump frequency, then

$$J = J_o \cdot \exp(-\Delta E_D / RT) \qquad (6.15)$$

The value of $\Delta E_D/RT$ can be estimated as per Eq. (6.16)

$$\Delta E_D / RT = C_1 / RT(C_2 + T - T_g) \qquad (6.16)$$

where C_1 and C_2 are WLF constants. This equation indicates the effect of glass transition temperature on solvent-induced crystallization. Further, the value of $\Delta F^*/RT$ can be estimated as per Eq. (6.17).

$$\frac{\Delta F^*}{RT} = \frac{4 \cdot \sigma_e \cdot \sigma_u \cdot T_m^2 \cdot b_o}{RT^2 \cdot \Delta H_u \cdot \Delta T} \qquad (6.7)$$

where σ_u and σ_e are growth surface and fold surface interfacial energies; ΔH_u, heat of fusion; T_m, melting point and ΔT supercooling. $T \cdot \Delta H_u \cdot \Delta T / T_m^2$ is the thermodynamic driving force. This shows that there is a reduction in the crystallite growth near the melting point due to decrease in thermodynamic driving force. Equation (6.14) can be influenced by the following factors.

1. The dilution of the polymer lowers the concentration of the crystallizing units, thereby reducing the jump frequency. This modifies Eq. (6.15) as Eq. (6.18)

$$J = v_2 \cdot J_o \cdot \exp(-\Delta E_D / RT) \tag{6.18}$$

where v_2 is the polymer volume fraction in the amorphous melt.

2. The diluent increases molecular mobility of the polymer/fibre and depresses glass transition temperature. The resultant glass transition temperature of the solvent-swollen polymer is shown in Eq. (6.19)

$$T_g = v_1 T_{g1} + v_2 T_{g2} \tag{6.19}$$

where T_{g1} is the glass transition temperature of the pure polymer; T_{g2} is the glass transition temperature of the diluents; v_1 volume fraction of the polymer and v_2 diluent volume fraction. This equation modifies the value of $\Delta E_D/RT$ shown in Eq. (6.16).

3. The diluent depresses the polymer of the fibre melting point.

$$\frac{1}{T_m} - \frac{1}{T_m^{\,o}} = \frac{R\upsilon_2}{\Delta H_u \upsilon_1}(v_1 - \chi v_1^2) \tag{6.20}$$

where $T_m^{\,o}$ is the melting point of the dry polymer of fibre; T_m, melting point of the polymer after dilution; υ_1 and υ_2 are molar volumes of the diluent and the polymer; χ interaction parameter. This equation modifies the value of $\Delta F^*/RT$ shown in Eq. (6.17).

4. The solvent alters the free energy barrier for the formation of a stable crystallizing interface.

$$\frac{\Delta F^*}{RT} = \frac{4 \cdot \sigma_e \cdot \sigma_u \cdot T_m^{\,2} \cdot b_o}{RT^2 \cdot \Delta H_u \cdot \Delta T} - \frac{2 \cdot \sigma_u \cdot T_m^{\,2} \cdot \ln v_2}{T \cdot \Delta H_u \cdot \Delta T} \tag{6.21}$$

The second term is an entropic contribution to the free energy barrier where units condense from a diluted melt. The diluent causes a widening of the crystallization temperature span and a drift of the span to the lower temperature than the pure polymer. This creates enhanced crystal growth in diluted system relative to the undiluted polymer at the same temperature because of a reduction of the activation energy. The assessment of G can be used to estimate or calculate any measurable crystallizable terms like crystallinity. If X denotes crystallinity, then

$$1 - [X_o(t) / X_o] = \exp(-A \cdot t^{\alpha}) \tag{6.22}$$

where $X_o(t)$ and X_o are instantaneous and final values of crystallinity. The constant A is dependent on the growth rate G. α is the Avrami exponent and it depends upon the nucleation process and crystallization geometry. For heterogeneous nucleated spherulites,

$$A = (4/3) \cdot \pi \cdot G^3 N_S / V_E \qquad (6.23)$$

where N_S/V_E is the nucleation density and $\alpha = 3$.

6.6.3 Strain-induced crystallization

The structure of any fibres or polymers is either isotropic semi crystalline, glassy amorphous, elastomeric or oriented semicrystalline. Plastics are either isotropic semicrystalline or glassy amorphous polymers. The fibres are basically oriented semicrystalline structure. When a semicrystalline polymer like fibre or plastic is subjected to deformation, it will cause a major structural change in the material. The deformation result in rapid extension and the crystallization is termed as strain-induced crystallization. Strain-induced crystallization is the mechanism that produces drawn fibre. During deformation or orientation process, orienting of the chain molecule leads to strain-induced crystallization. These oriented chains then act as nuclei for subsequent epitaxial crystallization. Sometimes, because of the nucleating action of oriented crystallized chains, the unoriented chains crystallize with chain folding. The orientation-induced crystallization has the following mechanism:

a. Chain alignment, i.e. stretching and/or orienting of the chain molecules of the melts, the flowing solutions or rubber-like material. This leads to formation of the nucleus.

b. Crystallization proceeds after the formation of nucleus in form of orienting chain molecules. This ensures growth of the crystallites.

The unoriented chains are present in random conformation. Strain-induced crystallization forms a shish-kebab structure. The extended chain structure first crystallizes out of the flowing solution and that these filament entities subsequently nucleate lamellae. In this process, long chains when stretched out are prone to form fibrils or fibrous crystals.

The extension is accomplished when fold chain blocks are tilted, sheared and eventually broken off the lamellae and become incorporated into microfibril. When the strain energy is sufficient, it will randomize chains at draw temperature and can create large local extensions. The chains would recrystallize during rapid extension, producing a row-nucleated structure.

When crystallization proceeds in the melt during elongation, one obtains a 'row nucleated' structure. The primary nuclei are the row nuclei. In row nucleation, the initial crystallites are the row nuclei. These are long needle like structures oriented in the strain direction and it is composed of extended chains. These long crystallites are oriented with long axes in the direction of the deformation. Crystallization then proceeds within a quiescent melt. This leads to formation of microfibrils with length of few microns and diameter 100–200 Å with tie molecules. Microfibrils are first formed in isolated deformation zones, and it propagates. These are basically needle-like crystallites and the properties of the fibre depend upon the characteristics of these fibrils, their interaction with the amorphous region and the state of the amorphous region. Mostly the state of the amorphous region is also an oriented state. The fibrillar crystallization is non-folded chain morphology – but a single pass through a crystallite by each chain. Fibrils have been referred to as strain – or stress-induced crystals.

The structure of fibre before drawing and after drawing as a result of deformation is radically different. The structures of undrawn or as spun fibres with minimum deformation will be lamellar crystallites at all orientations. Their thickness, i.e. long spacing is dependent on the crystallization temperature or supercooling during crystallization. Figure 6.4 is a schematic model portraying the structural changes at various stages of deformation with strain-induced crystallization. Figure (a) represents a crystalline-amorphous structural unit, which serves as basic unit for lamellar, spherulitic, and/or fibrillar morphologies. The structural unit is either isotropic or present in angular way with respect to the draw direction. The sample is macroscopically isotropic. This means that the orientation angle is equal in all directions or no orientation at all. The deformation or the strain-induced crystallization occurs in three steps. Those are: (i) pre-neck deformation of lamellar crystals, (ii) transformation of the lamellar structure into fibrillar structure and (iii) further deformation of the fibrillar structure.

Interlamellar amorphous layer induces the first stage of deformation with the extension of the amorphous region. The crystal lamella will slip past each other. This will result in minimum change in crystal orientation. The amorphous chain molecules may start to be aligned and strained without any appreciable change in orientation. The lamellae are all oriented after drawing. The thickness of the lamellae is also changed. When the fibre partially melts at the draw temperature, the large force present would result in rapid deformation of the amorphous chains. The conformational changes induce microfibril formation. When mobility of the chain molecules increases, there will be fracture of lamellae into small domains and subsequently slip past one another and align may happen when mobility increases.

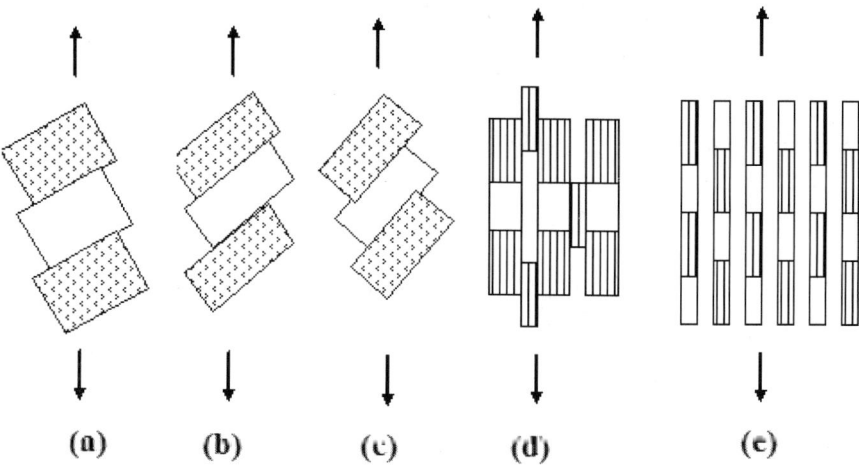

Figure 6.4   Schematic structural model of deformation, (a) undrawn and isotropic state, (b) lamellar sliding and expansion of amorphous state, (c) crystalline orientation and expansion of amorphous phase, (d) crystalline cleavage and amorphous orientation, (e) microfibril formation.

Higher deformation force results in extension of tie chains contained in both crystalline and amorphous regions (c). The deformation of lamellar structure involves rotation and sliding of stacked lamella, tilts the lamellae in addition to the lamellar slip. The crystalline orientation increases. Since the deformation of the tie molecules is accommodated by continued lamellar slip, the amorphous orientation factor may not increase and it will be at relatively lower value. This process brings the lamella into a stage where further deformation breaks them gradually into small folded chain blocks, which are incorporated into microfibrils. This breakup results in orientation of the crystallites (d). Continuous deformation results in the alignment of the amorphous chains to produce microfibrillar structure as shown in Fig. 6.4(e). Structure (d) and (e) occur at large deformations and it promotes a high degree of crystallite and amorphous orientation. The whole stack may be sheared yielding skewed microfibrils. This structural transformation is more important in as spun fibres. In such fibres solidification of a sufficiently strained melt starts by formation of highly oriented linear nuclei followed by epitaxial lateral growth of lamellae on such nuclei. The low deformation structures are associated with deformation and ultimate destruction of microspherulitic structure. The high deformation structures are associated with the transition to highly aligned fibrillar morphology.

6.6.4 Crystallization concurrent with chain growth

Crystallization concurrent with chain growth leads to formation of macroscopic polymer crystals. Such crystals are obtainable in especially favourable conditions. The monomers forming the macroscopic crystal can be joined up into chains by solid-state polymerization. By this method, specialty polymers with unique properties like metal-like conductivity or super conductivity at low temperature can be formed.

6.7 Crystallization of polyamide 6

The polymer melt is practically disordered molecular arrangement characteristic of a liquid. Quenching the melt results in a loosely packed amorphous structure and it is referred as amorphous I. This structure in the presence of moisture (it lowers T_g) or on being warmed stabilizes to a denser amorphous form, i.e. amorphous II presumably due to greater development of hydrogen bonding. Amorphous II is stabilized quenched material with closer molecular packing without any order. All these amorphous samples show different types of diffuse scattering in x-ray diffractogram. When the amorphous form is heated slightly above the T_g, short-range intermolecular order develops between chains without ordering along the length of the chains. The chains are mostly randomly twisted. These small volume elements consisting of an assembly of cylindrically shaped molecules having a unique spacing are formed. This results in γ-pseudohexagonal form. This is a nematic structure. The lateral order between chains results in a characteristic interchain spacing. There is no intermolecular order along the chains. Twist in chains results in statistical cylindrical symmetry along the chains. Structure exhibits birefrigent pattern different from that of normal spherulites.

If the chains are stretched moderately, longitudinal order of the amide groups is developed along the chain length, resulting in sharp diffraction effects in the x-ray pattern. This is β-hexagonal structure. The chain backbones are not a fixed configuration and may be partially twisted so that the amide groups make variable angles around the backbone chain. This produces a cylindrical projection and consequently hexagonal symmetry. The chains in this form may be parallel or antiparallel since their directionality is lost due to the twist. The hydrogen bonding in the structure is multidirectional between neighbouring chains, so that each chain is hydrogen bonded to each of its surrounding neighbours. This crystalline phase is only observed in oriented samples. The structure is defined by the three-dimensional ordering of the polar, i.e. amide groups along and among chains. This requires twist in each chain and result in shortening of the identity period along the chains and a statistical hexagonal distribution of angles of the amide groups around the chain.

If this β-structure is stretched to a higher degree, the twist in the molecular backbone is partially removed and the hydrogen bonds reform, so that there is no longer a hexagonal symmetry in the projection of the amide groups. This results in a monoclinic para crystalline α form. The oriented fibre does not exceed a density of 1.17 g/cm^3. There is a quantitative relationship between density and unit cell parameters of the α phase as a result of the sample history. Para crystalline phase is observed in oriented fibre samples. The structure is defined by the position of the fully extended C–C planer zigzag of the main chain. Chains are in planer configuration parallel to orientation direction, i.e. b axis. Planes can be in variable equilibrium positions between a and c axes which results in variable parameters of the monoclinic unit cell. Hydrogen bonding is restricted to sheet type formation. Relative position of polar groups between chains is not highly ordered. Heating this phase results in a pseudohexagonal form due to thermal motion which reforms to monoclinic on being cooled.

This highly ordered crystalline structure is defined by a three-dimensional ordering of the polar groups and fully oriented C–C planer zigzag of the main chains. This configuration requires antiparallel placement of neighbouring chains and forms close pack structure of nylon 6. Hydrogen bonding is in sheet form. The Bunn structure cannot be achieved in highly oriented morphology because it requires a fully antiparallel chains, the unit cell adopts the form, which results in maximum hydrogen bonding. This necessitates a partially disordered distribution of polar groups. The Bunn structure has the smallest unit cell and maximum density. The chains in this form are planner zigzag and the hydrogen bonds are made between adjacent chains forming a two-dimensional sheet type of structure. In this structure, both the polar groups and the backbone chains are in ordered arrangement. In order to accomplish this, the molecules must be in antiparallel position. The Bunn α structure is always found in crystalline bulk polymer, which shows spherulites morphology. Bunn α form is also associated with a chain folded morphology, which automatically results in antiparallel molecule arrangement.

The effect of heat is to bring a change in morphological structure of nylon in such a manner that there is change in the void content as well as relative size of the voids and their distribution. At a certain temperature, transition in the structure occurs, whereby the size of voids increases. A better segregation of crystalline and non-crystalline regions, resulting in easy movement of segments of chain in heat set PET as well as nylon above a certain temperature. The transition in the physical structure takes place in such a manner that the crystallites of ordered regions become enlarged and more perfect, at the same time there is better segregation of these regions from so-called non-crystalline or amorphous regions or voids.

6.8   Crystallization of poly (ethylene terephthalate)

Poly (ethylene terephthalate) can be converted into amorphous material by quenching the melt to room temperature. However, the amorphous material can be crystallized by heating above its glass transition temperature. Crystallization at comparatively lower temperature leads to formation of imperfect crystallites. At high temperature, the mobility of the segments is great. This permits rearrangements by segmental motion so that more crystallization occurs per incremental temperature. At high temperatures, the perfection of crystallites increased within very short times, due to partial melting and crystallization. Annealing at high temperature leads to the formation of large lamellar structures leaving only a small portion of the original fraction in the amorphous region. The lamellar structure is absent at temperature at 100°C. The stackings of lamellae are visible when crystallized at high temperature of annealing, when the dimensions of the lamella increases, the number of corresponding interlamellar links becomes less. The increase in crystallinity during the process of crystallization generally occurs in two ways. There will be (1) increase in total number of crystallites and (2) enhancement of average perfection of the existing crystallites. The former can occur by nucleation and formation of newer crystallites. This may result in smaller crystallites. Even sometimes the size of crystallites reduces to give way for more crystallites. The latter can be formed when there will be more movement or fluidity of the molecular chains, so that the defects within the crystalline region can be dissipated. This requires high temperatures as discussed previously.

The isomerization rate is accelerated at high temperature. The trans isomer followed by the gauche isomers of PET in amorphous region converts and increases the trans isomers of PET in crystalline region. The amorphous trans isomers are associated with the extended chain units. It makes up the interlamellar links, although the amorphous trans isomer fraction is relatively small. There is presence of intercrystalline links between the thin lateral edges of chain folded lamellar crystals in spherulitic polymers. The decrease in amorphous trans content may correspond to a decrease of the number of tie molecules.

6.9   Melting

Melting is described as the first order of transition, where the material is converted from the solid state to the liquid state or vice versa. Structurally a molten polymer is devoid of any secondary or intermolecular forces. On the other hand, a semicrystalline polymer or fibre consists of intermolecular forces both in amorphous and crystalline region. At glass transition temperature, the intermolecular forces are completely eliminated from the amorphous region.

With heating above glass transition temperature, the intermolecular forces gradually eliminated from the crystalline region. Melting is considered as that point where all the intermolecular forces eliminated or destroyed by heating and crystallinity reduces to zero.

The crystalline texture of any polymer is divided into three regions corresponding to the degree of thermal motion of molecules. (1) The molecules, which are capable of micro-Brownian motion above its glass transition temperature, (2) the molecules, which are only capable of the local twisting motion but incapable of diffusional motion even above glass transition. Such region is present within the lamellae and (3) the crystalline region with perfect regularity and the thermal motion associated with this region will be initiated at the temperature region of the crystalline absorption.

The melting temperature of any polymer and/or fibre depends on the energy of intermolecular interaction, i.e. cohesive energy and on the internal mobility of its molecules, i.e. their ability to undergo conformational changes. A crystal melts when the amplitude of vibration of its atoms or molecules reaches a critical value determined by the distance between neighbouring molecules. It increases with temperature. The critical value occurs at the melting point. The greater is the cohesive energy and smaller the internal mobility, the higher will be the melting temperature. The internal mobility of the molecules is more controlled by potential energy barriers for rotation. Higher energy requires for the rotation restricts the internal mobility and it leads to higher melting point. On the other hand, higher mobility leads to higher chain flexibility.

Melting temperature of some of the polymers is shown in Table 6.1. Polyethylene and polypropylene are made up of flexible macromolecules with very weak intermolecular forces. As a result of this, the melting point of polyethylene is 135°C and polypropylene 165°C. Aliphatic polyester also melts at lower temperature because of chain flexibility. The CH_2 (methylene) groups in this polymer is turned around 80° around the ester group. This also facilitates rotation about the direction of the bond thus resulting a lower melting temperature. In general, lower melting temperature can be due to increased chain flexibility and may be due to a large number of conformations forming in the molten state. Benzene or aromatic group restricts chain mobility and it enhances the melting point. In case of poly (ethylene terephthalate), the ester group as well as the methylene group is present in a plane and restricts rotation of the bond as well as number of conformations. As a result of this, and due to presence of benzene ring poly (ethylene terephthalate) shows higher chain stiffness and thus higher melting point.

Further, when amide groups replace ester group in any polymer, chain flexibility decreases and cohesive energy density sharply increases. This results in higher melting point for polyamide in comparison with polyesters.

So polyamide 6 and polyamide 66, though contain flexible molecular chains, still exhibit higher melting points. Polyurethanes contain both amide and ester groups. This reduces cohesive energy density. Polyurethane shows an intermediate melting point in-between polyamide and polyester.

The chain flexibility also differs in different isomers. Depending upon structure, some polymers have different isomers like ortho-, meta- or para-isomers. The number of conformations in para molecules is not large and the chain is comparatively rigid. On the other hand, the number of conformations in the meta-isomer is higher than the para-isomers. Owing to this, the para-isomer of any polymer melts at a lower temperature than other isomers. This behaviour can be observed for the polymers like polybutadiene, poly phenylenes, polyesters and aromatic polyamides.

The series polyesters, polyamides and polyurethanes display differences in melting points with odd and even numbers of carbon atoms. The polymers with even number of carbon atoms in their monomers like dicarboxylic acids, dihydric alcohols or diamines have higher melting points than those with odd numbers because of higher cohesive density. The thermal vibrations of molecules in crystals differ depending on the evenness or oddness of the carbon atom present in the backbone.

Table 6.1 Melting temperature of selected polymers

No	Polymer	T_m (°C)
1	Polyethylene	137
2	Polypropylene (iso)	176
3	Polyvinyl chloride	227
4	Polystyrene	240
5	Polyamide 6	214
6	Polyamide 6,10	
7	Polyamide 6,6	260
8	Polyethylene adipate	50
9	Polybutylene terephthlate	220
10	Polyethylene terephthlate	265

A polymer or the fibre does not have any definite melting point. It melts within a certain temperature interval, the breadth of which depends on the history of the sample. The final melting temperature corresponds to the final disappearance of the crystallinity. This forms the upper limit of the melting range. The lower limit of the melting range is the crystallization temperature. So the melting consists of a broad temperature range. The shape of the melting curve between these two extremes reflects the shape of the crystallite stability present. A sharp transition from the crystalline to the molten state may be observed only in case of perfect order in crystalline phase, larger or bigger crystals and/or higher crystallinity. So melting provides information on the nature of the crystalline structure being destroyed. Melting of the crystals at any temperature below the melting range means that there is only partial melting, where the surface structure will be melted with more micro-Brownian motion.

The fibre or the polymer consists of different types of crystallites with different sizes, perfection and stability. These types of crystallites melt at different temperature. So for the polymers or fibres, melting is a progressive process, extending ever a wide range of temperature. The more defective crystals or the smaller crystals will be melting first. The more perfect and stable crystals will melt at higher temperatures. Small average size of the crystallites or less perfect surface regions will have large surface to volume ratio and it will result in a depression of the melting point. On the other hand, the increase of the melting point can be attributed to crystal thickening, crystal perfection and fold surface smoothing of the crystalline region. The melting point and stability of the crystals can be related by means of Eq. (6.24) and it is known as Thomson–Gibb's equation.

$$T_m = T_m^{\ 0}\,[1 - 2\,(\sigma_e/\Delta H_f l_c)] \qquad\qquad (6.24)$$

where T_m is the melting point of the material; $T_m^{\ 0}$, melting point of an infinite perfect crystal; σ_e, surface energy of the crystal; ΔH_f, heat of fusion of the material and l_c, thickness of the crystalline lamella. The value of the melting point as well as the melting range is influenced by crystallite size, fold surface structure and the crystallite imperfection. A mean thickness of the crystalline lamella (l_c) can be calculated as $l_c = L.\ X_{DSC}$, where L is the mean long spacing as obtained by applying Bragg's law to the peak position in SAXS pattern.

During crystallization process, the crystal perfection occurs at lower rate of heating to give more scope for reorganization of the chain molecules. The crystal thickening or the transformation implies melting of the original crystallites with subsequent nucleation at higher temperature, crystal thickening and crystal perfection. During slow heating, the new crystalline material develops at the fold surface of the existing crystals at the expense of the neighbouring amorphous layers. As a consequence, crystal thickening and fold surface smoothing can occur. The melting point can be influenced by symmetry of the polymer, tacticity, branching and/or intermolecular bonding. Molecular weight, co-polymerization and presence of solvent or plasticization can also influence the melting behaviour as per Eqs. (6.25)–(6.27).

(a) Molecular weight

$$\frac{1}{T_m} - \frac{1}{T_m^{\;0}} = \frac{2 \cdot R \cdot M_o}{\Delta H_u \cdot \bar{M}_n} \tag{6.25}$$

(b) Co-polymerization

$$\frac{1}{T_m} - \frac{1}{T_m^{\;0}} = -\frac{R \cdot \ln X_A}{\Delta H_u} \tag{6.26}$$

(c) Solvent and plasticization

$$\frac{1}{T_m} - \frac{1}{T_m^{\;0}} = \frac{R \cdot V_u}{\Delta H_u \cdot V_1}(\varphi_1 - \chi \cdot \varphi_1^{\;2}) \tag{6.27}$$

where M_o is the molecular weight of the monomeric unit; $\bar{M}_n$, average number of molecular weight; X_A, molecular fraction of comonomers A; V_u, molar volume of the polymer repeat unit; V_1, molar volume of the solvent; ϕ_1, volume fraction of the solvent and χ interaction parameter. The interaction parameter if negative is good for solvent interaction with the polymer.

Blending of polymer also influences the melting points of each of the components. Basically, there will be depression in melting point due to disruption of the crystals as per Eq. (6.28).

$$1/T_m - 1/T_m^{\;0} = [R \cdot V_{2u} / H_{fu} \cdot V_{1u}](\chi_{12} \cdot V_1^{\;2}) \tag{6.28}$$

where 1, 2 stand for blend components, 1 second component, H, molar heat of fusion and V, molar volume. The polymer–polymer interaction parameter (χ_{12}) can be calculated from the depression of Eq. (6.28) or (6.29).

$$1/T_m = -[BV_{2u}V_1^{\;2} / H_{fu}T_m] + 1/T_m^{\;0} \tag{6.29}$$

where $\chi_{12} = B \cdot V_{1u}/R \cdot T_m$ and its value can be obtained by plotting $10^3/T_m$ (K^{-1}) versus $V_1^{\;2} \cdot 10^5/\cdot T_m$ (K^{-1}) and the slope will be $= BV_{2u}/H_{fu}$.

6.10 Thermal methods of analysis

Melting behaviour, melting points, melting range, crystallization temperature as well as crystallization range can be obtained in a Differential Thermal Analysis (DTA) or Differential Scanning Calorimetry (DSC). The nomenclature committee of the International Confederation for Thermal Analysis (ICTA) has defined Differential Thermal Analysis (DTA) and Differential Scanning Calorimetry (DSC) as follows:

(1) DTA: A technique in which the temperature difference between a substance and a reference material is measured as a function of

temperature whilst the substance and reference material are subjected to controlled temperature programme. The record is differential thermal or DTA curve. The temperature difference should be plotted on the ordinate with endothermic reactions downwards and temperature (T) or time (t) on the abscissa increasing from left to right.

(2) DSC: A technique in which the difference in energy inputs in a substance and a reference material is measured as a function of temperature whilst the substance and reference material are subjected to controlled temperature program.

DTA and/or DSC measures the heat/energy change occurring in a substance as a function of temperature. Experimentally the sample is heated side by side with an inert reference material at a uniform rate and the temperature/energy/power difference between them is measured as a function of temperature. The resulting curve is called a 'thermogram' from which information can be derived. The main advantage of DTA/DSC is that it is simple, rapid and subject to continuous recording.

The different information available from DTA/DSC may be classified into three basic types: endothermic with heat absorption, exothermic with heat evolution and second order transition where the specific heat undergoes a sudden change. Transformations in high polymers can be divided into those involving physical transitions and those involving chemical reactions. The physical transitions are glass transition, melting, crystallization, cold crystallization, crystal–crystal transition, crystal disorientation and heat of reaction. A schematic diagram of a thermogram obtained from DTA or DSC is shown in Fig. 6.5

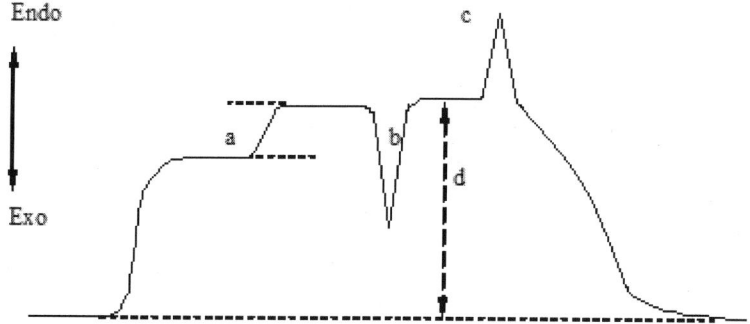

Figure 6.5   Schematic figure of a DTA/DSC thermogram: a = second-order transition, b = crystallization peak, c = fusion peak and d = deflection directly proportional to C_p

Further readings

1. Claudio De Rosa and Finizia Auriemma, *Crystals and Crystallinity in Polymers: Diffraction Analysis of Ordered and Disordered Crystals,* Wiley, 2013.

2. *Encyclopedia of Polymer Science and Engineering*, Wiley, New York, 1986.

3. L. Mandelkern, *Crystallization of Polymers*, Mc-Graw Hill, New York, 1964.

4. M. Lewin and E. M. Pearce, *Handbook of Fibre Science & Technology*, Marcel Dekker, New York, 1985.

5. H. Tadokoro, *Structure of Crystalline Polymers*, Wiley, New York, 1979.

6. Ewa Piorkowska and Gregory C. Rutledge, *Handbook of Polymer Crystallization,,* Wiley, 2013.

7.1 Introduction

The fibre forming process from polymer melt or solution consists of deformation and it results in orientation of polymer molecular chains. Structurally the fibre consists of crystalline regions and amorphous regions with different molecular arrangements. Owing to this, orientation in fibres can be of different types and it is specified to different regions or components. These different orientations are as follows:

(a) Orientation of the chain molecules,

(b) Average orientation of the chain molecules,

(c) Orientation of the chain molecules in the amorphous phase,

(d) Orientation of the crystallographic axes of the crystallites,

(e) Orientation of the crystallites.

Chain orientation is a phenomenon unique to the polymers only and it refers to the alignment of the chain molecules. Polymers are basically consisting of long molecular chains, having strong covalent bonds along the chain axis and weaker secondary forces in the transverse direction. This creates anisotropic character in the chain molecule. Random molecular chains are isotropic as the covalent bonds are present in all directions. Orientation of the chain molecules is the result of deformation. A given strain increases the average end-to-end distance with higher angle between the chain axes [Eq. (2.25)]. The mathematical derivation of the end-to-end distance is shown in Section 2.7.1. As the molecular chains in the polymer or the fibre are random and they differ from crystalline region and amorphous region, it is always preferable to quantify the orientation of the chain molecules as the average orientation of the chain molecules. Average orientation of the chain molecules is related to the weighted average orientation of the chain molecules present in different phases.

7.2 Orientation factor

Orientation is best expressed quantitatively by means of 'Hermans orientation function'. This function was first derived by the Dutch scientist P. H. Hermans in 1946 and so termed as Hermans orientation function or simply as orientation factor. This factor may conveniently be connected with an average angle of orientation of the elementary particle. This is also related to the optical properties of the material with segmental orientation. A brief description of orientation factor or orientation function is discussed below.

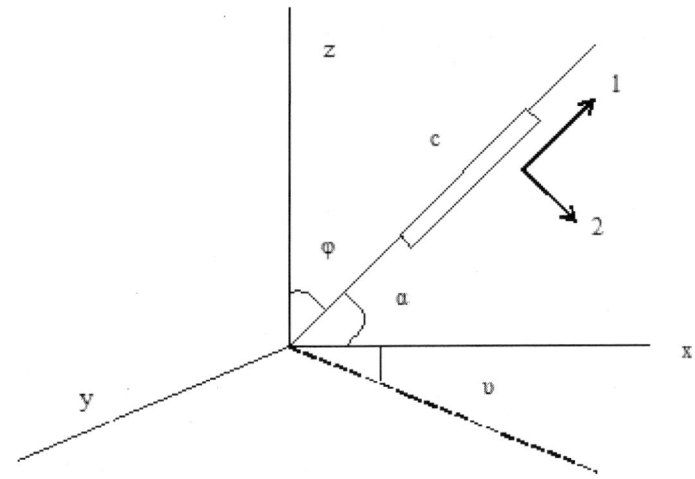

Figure 7.1   Polymer chain segment (*c*) with coordinate system.

The orientation of the chain segment with respect to *x*, *y*, *z* co-ordinate system is shown in Fig. 7.1. The polarizability associated for each chain segment can be expressed by a component parallel to the chain axis (p_1) and a component perpendicular to the chain axis (p_2). The directional vectors of the chain segment are mentioned as 1 and 2 in the figure. It is assumed that the electrical vector of the propagating light is along *z* axis. The electrical filed parallel to the chain axis $E_z \cos \phi$ and the polarization along the chain segment is given by $p_1 E_z \cos \phi$. The contribution of this polarization to the polarization along the *z* axis amounts to $p_1 E_z \cos^2 \phi$. The contribution to polarization along *z* from transverse polarization of the segment equals to $p_2 E_z \sin^2 \phi$. The total polarization along the *z* axis caused by electrical vector in the *z* direction can be expressed by the term P_{zz}. The value of P_{zz} is equal to the sum of these two contributions and shown in Eq. (7.1).

$$P_{zz} = p_1 E_z \cos^2 \phi + p_2 E_z \sin^2 \phi = E_z (p_1 \cos^2 \phi + p_2 \sin^2 \phi) \qquad (7.1)$$

The polarizability in the *z* direction can be termed as p_{zz} and its value can be as per Eq. (7.2)

$$p_{zz} = P_{zz}/E_z = (p_1 \cos^2 \phi + p_2 \sin^2 \phi) \tag{7.2}$$

In a similar way, the polarizability in the x direction can be termed as p_{xx} and its value can be as per Eq. (7.3)

$$P_{xx} = p_1 \cos^2 \alpha + p_2 \sin^2 \alpha \tag{7.3}$$

Since $\cos \alpha = \cos v \cdot \sin \phi$, the equation can be modified as Eq. (7.4)

$$P_{xx} = P_2 + (p_1 - p_2) \sin^2 \phi \cos^2 v \tag{7.4}$$

The polarizability tensor (p_{ij}) can be converted to the corresponding refractive index tensor (n_{ij}) as per the Lorentz–Lorentz equation and it is shown in Eq. (7.5)

$$\frac{n_{ij}^2 - 1}{n_{ij}^2 + 2} = -\frac{4\pi}{3} p_{ij} \tag{7.5}$$

Inserting Eqs. (7.3) and (7.4) in Eq. (7.5) yields Eqs. (7.6) and (7.7)

$$\frac{n_{zz}^2 - 1}{n_{zz}^2 + 2} = \frac{4\pi}{3} (p_1 \cos^2 \phi + p_2 \sin^2 \phi) \tag{7.6}$$

$$\frac{n_{xx}^2 - 1}{n_{xx}^2 + 2} = \frac{4\pi}{3} \{p_2 + (p_1 - p_2) \sin^2 \phi \cos^2 v\} \tag{7.7}$$

The quantitative definition of the birefringence and the average refractive index is given in Eqs. (7.8) and (7.9).

$$\Delta n = n_{zz} - n_{xx} \tag{7.8}$$

$$|n| = \tfrac{1}{2}(n_{xx} + n_{zz}) \tag{7.9}$$

Combination of Eqs. (7.8) and (7.9) yields the expression of n_{xx} and n_{zz} and it is shown in Eqs. (7.10) and (7.11)

$$n_{xx} = |n| - (\Delta n/2) \tag{7.10}$$

$$n_{zz} = |n| + (\Delta n/2) \tag{7.11}$$

And deducting and comparing of Eqs. (7.6) and (7.7) yield Eqs. (7.12) and (7.13)

$$\frac{n_{zz}^2 - 1}{n_{zz}^2 + 2} - \frac{n_{xx}^2 - 1}{n_{xx}^2 + 2} = \frac{6|n|\Delta n}{(n_{zz}^2 + 2)(n_{xx}^2 + 2)} \approx \frac{6|n|\Delta n}{(|n|^2 + 2)^2} \tag{7.12}$$

$$\frac{n_{zz}^2 - 1}{n_{zz}^2 + 2} - \frac{n_{xx}^2 - 1}{n_{xx}^2 + 2}$$

$$= \frac{4\pi}{3} [(p_1 \cos^2 \varphi + p_2 \sin^2 \varphi) - \{p_2 + (p_1 - p_2) \sin^2 \varphi \cos^2 v\}] \tag{7.13}$$

or $\qquad \dfrac{n_{zz}^{2}-1}{n_{zz}^{2}+2}-\dfrac{n_{xx}^{2}-1}{n_{xx}^{2}+2}=\dfrac{4\pi}{3}[(p_{1}-p_{2})(1-\sin^{2}\varphi-\sin^{2}\varphi\cos^{2}\upsilon)]$ $\qquad$ (7.14)

Equating Eqs. (7.12) and (7.14), the value of birefringence is expressed in Eq. (7.15)

$$\Delta n = \frac{(|n|^{2}+2)^{2}}{6|n|} = \frac{4\pi}{3}[(p_{1}-p_{2})(1-\sin^{2}\varphi-\sin^{2}\varphi\cos^{2}\upsilon)] \qquad (7.15)$$

Orientation is uniaxial and all angles υ are equally probable. The average square of the cosine of υ may be calculated from Eq. (7.16).

$$|\cos^{2}\upsilon| = \frac{1}{2\pi}\int_{0}^{2\pi}\cos^{2}\upsilon = \frac{1}{2} \qquad (7.16)$$

Insertion of Eq. (7.16) in Eq. (7.15) and averaging for all segments over all θ angles give Eq. (7.17)

$$\Delta n = \frac{(|n|^{2}+2)^{2}}{6|n|}\frac{4\pi}{3}(p_{1}-p_{2})\left|1-\frac{3|\sin^{2}\varphi|}{2}\right| \qquad (7.17)$$

If the maximum intrinsic birefringence in Eq. (7.17) is given as in Eq. (7.18)

$$\Delta n_{0} = \frac{(|n|^{2}+2)^{2}}{6|n|^{2}}\frac{4\pi}{3}(p_{1}-p_{2}) \qquad (7.18)$$

Then Eq. (7.17) can be modified and shown in Eq. (7.19)

$$\Delta n = \Delta n_{0}\left|1-\frac{3|\sin^{2}\varphi|}{2}\right| \qquad (7.19)$$

As per Eq. (7.19), Hermans orientation function (f) is defined as

$$f = \frac{\Delta n}{\Delta n_{0}} = \left|1-\frac{3|\sin^{2}\varphi|}{2}\right| = \left|\frac{3|\cos^{2}\varphi|-1}{2}\right| \qquad (7.20)$$

The maximum orientation will have $\phi = 0°$ or $\sin^{2}\phi = 0$ and this makes the orientation function value as 1 ($f = 1$). This will indicate for a system where complete orientation parallel to the fibre axis will be present. On the other hand, for a random orientation $\phi = 90°$ or $\sin^{2}\phi = 0$ and this makes the value of f as $-\frac{1}{2}(f = -\frac{1}{2})$. This will indicate the system where the orientation is perpendicular to the fibre axis. Non-oriented system will have $f = 0$.

In a fibre, the main axis of the collective particles, i.e. monomeric residues or crystallites form an angle, ϕ, with the fibre axis. The fibre of uniform orientation is related to the situation, where all particles being oriented at an angle, ϕ, to the fibre axis. The average angle of orientation in a fibre is then equal to the angle of orientation, ϕ, of that fibre with uniform orientation. Then, orientation factor can be expressed as per Eq. (7.19).

The orientation in any fibre with a single crystal and ideal lattice arrangement of the molecules represent the condition of maximum orientation of the object. The order in the crystal is characterized not merely by the parallel alignment of the chain axis of the unit cells (longitudinal orientation) but also by an ideal orderliness in direction perpendicular to them (lateral orientation). The maximum orientation of the fibre is the state of ideal longitudinal orientation with the longitudinal axes of all chain molecules or crystalline regions are parallel to the fibre axis. In this case, lateral orientation may still be random. The state of orientation in a fibrous object can be described by suitably indicating the spatial arrangement of the longitudinal axis of all monomeric residues.

7.3 Crystallite orientation

In a polycrystalline material, the fibre texture can be conveniently expressed in terms of $(\cos^2 x)_{av}$, where x is the angle between a chosen crystallographic direction and a reference direction in the sample. $(\cos^2 x)_{av}$ can also be determined, even if there are no reflecting planes normal to the chosen crystallographic direction. When the desired crystallographic direction coincides with the z axis of a Cartesian co-ordinate system, the crystal may be fixed at any arbitrary position with respect to the x and y axes. Let a unit vector, N, along the normal to a reflecting set of planes can be represented as

$$N = e \cdot i + f \cdot j + g \cdot k \qquad (7.21)$$

where e, f, g are the direction cosines of N with respect to x, y, z, respectively. Let Q be a unit vector along the reference direction in the sample and φ be the angle between Q and N. Then $\cos \varphi = N \cdot Q$ and

$$
\begin{aligned}
(\cos^2 \varphi)_{av} = {} & e^2 (\cos^2 x)_{av} + f^2 (\cos^2 y)_{av} \\
& + g^2 (\cos^2 z)_{av} + 2ef \{(\cos x)(\cos y)\}_{av} \\
& + 2fg \{(\cos y)(\cos x)\}_{av} + 2ge\{(\cos x)(\cos z)\}_{av}
\end{aligned} \qquad (7.22)
$$

Since $Q \cdot \phi = \cos z$, the quantity $(\cos^2 \varphi)_{av}$ for the planes (hkl) can be experimentally determined from the distribution of diffracted intensity from these planes; the co-efficient e, f and g can also be calculated from the knowledge of the crystal structure. Hence the above equation contains six unknown averages. The quantity $(\cos^2 x)_{av}$ can therefore be evaluated by solving six simultaneous equations involving these six unknowns. The equation can be evaluated for five set of reflecting planes having normals in different directions from each other would provide five of the required equations. The sixth equation is the orthogonality relation and is:

$$(\cos^2 x)_{av} + (\cos^2 y)_{av} + (\cos^2 z)_{av} = 1 \qquad (7.23)$$

In practice, the number of equations required might be reduced considerably. For example, if the orientation of the crystal C axis is being determined, then for the reflection of the type (hk0), at least two of the cross product can be made to vanish since $q = 0$. Therefore, not more than three reflections of this type would require a solution for $(\cos^2 x)_{av}$. In the trivial case, where a set of reflecting planes is normal to the crystallographic direction, the co-efficient $e = f = 0$ and $g = 1$, and so:

$$(\cos^2 \phi)_{av} = (\cos^2 x)_{av} \qquad (7.24)$$

Orientation can then be determined from a single (hkl) plane and it is either (h00) or (0k0) or (00l) plane. Herman's orientation factor as defined in Eq. (7.19) can be calculated as per the following equation

$$f_{hkl,z} = \frac{1}{2} \cdot [3(\cos^2 \phi)_{hkl,z} - 1] \qquad (7.25)$$

$$\cos \varphi\phi = \cos \theta \cdot \cos \rho \qquad (7.26)$$

$$(\cos^2 \phi)_{hkl,z} = \frac{\displaystyle\int_0^{\pi/2} I_{hkl}(\phi) \cdot \sin \phi \cdot \cos^2 \phi \cdot d\phi}{\displaystyle\int_0^{\pi/2} I_{hkl}(\phi) \cdot \sin \phi \cdot d\phi} \qquad (7.27)$$

where $I_{hkl}(\phi)$ is the scattered intensity of the (hkl) reflection at the azimuthal angle ϕ.

For random orientation,

$$(\cos^2 \phi)_{hkl,x} = (\cos^2 \phi)_{hkl,y} = (\cos^2 \phi)_{hkl,z} = 1/3 \qquad (7.28)$$

$$(\cos^2 \phi)_{hkl,x} + (\cos^2 \phi)_{hkl,y} + (\cos^2 \phi)_{hkl,z} = 1 \qquad (7.29)$$

or

$$f_{a,j} + f_{b,j} + f_{c,j} = 0 \qquad (7.30)$$

7.4   Limiting cases of orientation

There are two extreme boundary cases of orientation, when considering the quantitative data relating to the connection between optical properties and orientation of fibres.

(i)   No orientation at all: This includes isotropic objects. The distribution of the monomeric residues in the amorphous portions and of the crystallites is such that their anisotropies cancel each other out on the average.

(ii) Ideal orientation: In this case, the ideal longitudinal (uniaxial) orientation is merely supposed that the longitudinal axes of all the crystallites lie parallel to the fibre axis. An exceptional case of ideal orientation is such that the fibre displays ideal orientation and the density of the fibre is the density of the crystalline areas. Such a fibre is known as an 'ideal fibre'. Here the optical constants depend upon the average density of the packing and not upon the quantitative distribution between the crystalline and amorphous substance.

7.5   Measurement of orientation

Orientation can be determined by a number of methods like optical methods measuring birefringence, wide angle diffraction for crystallite orientation, small angle scattering for lamellar, infrared spectroscopy, Raman spectroscopy, nuclear magnetic resonance spectroscopy, sonic modulus. The principle of measurement of orientation is discussed briefly. However, detail of the process is discussed in individual chapters.

7.5.1 Birefringence

Birefringence measurement is a simple and classical method of investigation of orientation. The Herman's orientation function is proportional to the birefringence as per Eq. (7.31)

$$f = \Delta n / \Delta n_0 \qquad (7.31)$$

where Δn_0 is the maximum birefringence. The value can be negative or positive values. Polymer with polarizable units in the main chain have positive Δn_0 values, whereas polymers with strongly polarizable pendant groups show negative values. Birefringent values are positive for polymers or fibres like polyethylene or polyethylene terephthalate. On the other hand, polystyrene or polyacrylonitrile shows Δn_0 negative values. The birefringence value is the result of several components and it can be shown as follows.

$$\Delta n = \Delta n_f + \Delta n_d + \Delta n_c + \Delta n_a \qquad (7.32)$$

where Δn_f is the form birefringence, Δn_d, deformation birefringence, Δn_c, orientation induced birefringence originating from the crystalline components and Δn_a orientation induced birefringence originating from the amorphous components. Form birefringence occurs only in multiphase systems, where the anisotropic system consists of different components with different refractive systems. Deformation birefringence occurs because of the stresses in bonds due to deformation and so it is referred as deformation birefringence. However, this deformation birefringence is insignificant in any system.

For uniaxial orientation system, both form birefringence and deformation birefringence can be neglected and the equation can be modified and shown in Eq. (7.33)

$$\Delta n = \Delta n_c + \Delta n_a \qquad (7.33)$$

Birefringence is basically an aggregated property. The results always weighted towards high polarizability groups like benzene rings. This also assumes additivity schemes for polarizability and it might be uncertain for C–C bonds. This equation can be modified by taking the fraction of crystalline and amorphous components into consideration. The modified equation is mentioned in equation

$$\Delta n = \Delta n_0 [w_c \cdot f_c + (1 - w_c) f_a] \qquad (7.34)$$

where f_c and f_a are the Hermans orientation functions of the crystalline and the amorphous components and w_c is the fraction of crystallinity of the system.

Birefringence measurements can be made in a polarized light microscope by any of the methods like (a) refractive index method, (b) compensator method and (c) fringe method. The refractive method is a classical method. The birefringence can be determined as per Eq. (7.8) by directly measuring two refractive indices. The two refractive indices can be measured by immersing the fibre in an immersion fluid that matches the refractive index of the fibre. However, this method is more tedious and the matching fluid is often toxic and objectionable for use. So this method is not used at present for birefringent analysis.

The configuration in a polarizing microscope with the analyser and polarizer for analysis by compensator method is shown in Fig. 7.2. Analyser and polarizer should be crossed and should be oriented at an angle 45° to the main optical axes of the sample. A compensator is a made of birefringent material and it is introduced between the sample and analyser. The compensator makes it possible to change the optical retardation introduced by the sample is compensated for. There are different types of compensators. The Babinet and the tilt compensators are probably the most common compensators used for the measurement.

The Babinet compensator consists of two quartz wedges cut so that their optical axes are mutually perpendicular. A longitudinal shift of the lower wedge changes the optical retardation. A series of fringes is observed around the centre provided that monochromatic light has been used. The longitudinal shift of the lower wedge which shifts the zero-order fringes to the centre gives, after suitable calibration, of the sample. With a light source, containing a continuous spectrum of energy in wavelengths from 400 nm to 750 nm, coloured fringes are observed on both sides of the central black fringe with an optically isotropic sample. The optical retardation is wave length dependent and the complex subtraction of light of different wavelengths means that each optical retardation is characterized by specific colour. This colour scale, i.e. colour as a function of optical retardation is given in a chart and known as Michel–Levy chart. The tilt compensator consists of either a single birefringent calcite plate (Berek type) or two plates of either calcite or quartz cemented together (Ehringhaus compensator). The schematic diagram of a Babinet compensator consisting of two quartz wedges is shown in Fig. 7.2.

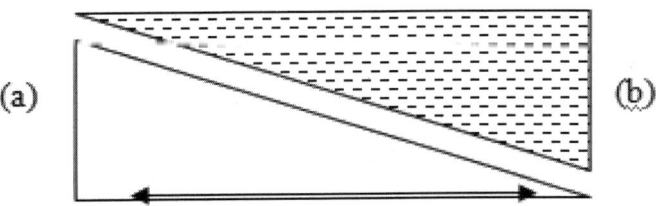

Figure 7.2   Babinet compensator consisting of two quartz wedges, where (a) is moveable and (b) is fixed.

The change in optical retardation is achieved by rotation of the plate, which causes a change in both absolute travelling length and refractive index. The birefringence is calculated from the measured optical retardation according to the following equation

$$\Delta n = R \cdot \lambda / d \qquad (7.35)$$

where R is the measured optical retardation, λ, the optical wavelength in microns, and d is the sample thickness. The compensator method is ideal for static measurement. Fringe method is a modification of the compensator method. In this method, the signal emerging from the analyser is used to be dispersed by a front surface grating and the projected spectrum is used to be photographed. The analyser is then removed and the optical path and replaced by the calibrated quartz plate with its attached polarizing analyser. The quartz system is also dispersed by grating and photographed. The photographic fringe pattern of the fibre and the quartz is compared for birefringence calculation as per the equation mentioned in Eq. (7.36)

$$\Delta n = \Delta f_F d_Q \Delta n_Q / \Delta f_Q \cdot d_F \qquad (7.36)$$

where Δf is the fringe interval count, d, the thickness and the subscript F stands for the fibre and Q for the quartz. For dynamic measurements, it is possible to

remove the compensator and to measure the intensity of the transmitted light
(*I*) from which optical retardation can be calculated.

$$I = I_0 \cdot \sin^2(\pi \cdot R) \tag{7.37}$$

where I_0 is the intensity of the incoming light. This equation is applicable to
samples showing an optical retardation less than the wavelength of the light.
The in-plane birefringence can be calculated according to Eq. (7.35)

Optically clear amorphous polymers are readily studied also using thick
samples. However, semicrystalline polymers scatter light and the method of
using visible light is only applicable to the study of thinner samples (<100 μm).
For proper experiments transparent specimen and sectioning technique may
be required.

7.5.2 X-ray diffraction

X-ray diffraction is used for the analysis of crystalline structure of any sample
including fibres and polymers. The orientation from x-ray is the orientation
of the crystallites. The orientation of (*hkl*) planes in semicrystalline polymers
is revealed by wide angle x-ray diffraction. Figure 7.3 shows the diffraction
patterns of two samples of the same polymer. The unoriented polymer shows
concentric rings for the intrachain (001) and interchain (*hk*0) reflections. The
oriented polymer shows diffractions of both (001) and (*hk*0) and it is dependent
upon the azimuthal angle ϕ. The intrachain diffractions concentrate on the
meridional and the interchain reflections are concentrated equator.

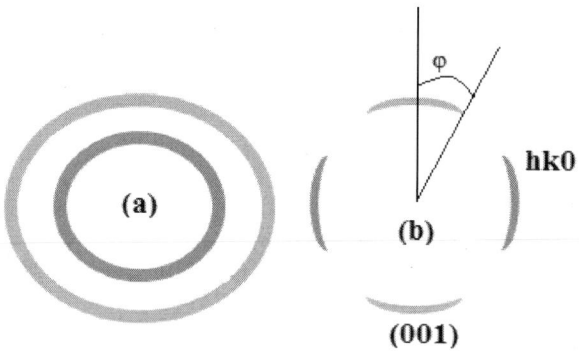

Figure 7.3   XRD of (a) powdered and (b) oriented sample photograph.

Figure 7.3 shows the diffraction patterns of two samples of the same polymer.
The unoriented polymer shows concentric rings for the intrachain (001) and
interchain (*hk*0) reflections. The oriented polymer shows diffractions of both
(001) and (*hk*0), which are strongly dependent on the azimuthal angle (ϕ).

The intrachain reflection is concentrated to the meridian and the interchain reflection to the equator.

The orientation of the crystallites to the fibre axis can be derived from the intensity distribution of the meridional x-ray reflections. These reflections are, however, usually weak and under certain conditions absent. This orientation factor could be derived from a knowledge of the average orientation of the planes at right angles to the meridional ones, i.e. from the equatorial x-ray reflections produced by planes oriented parallel to the fibre axis on drawing. These planes are preferable in two other aspects: the correction factor, for the equatorial reflections, with finite change in Bragg angle is negligible and furthermore the Lorentz factor may be ignored.

The orientation expressed by the average cosine square of the angle ϕ of the diffracting planes $(hk0)$ and (001) can be calculated from Eq. (7.27). The orientation of the different crystal planes (100), (010) and (001) of an orthorhombic cell to a common director is shown in Eq. (7.38) and this equation may be applied to the simple case with only one interchain equation, as shown in Eq. (7.39)

$$|\cos^2 \phi_{100}| + |\cos^2 \phi_{010}| + |\cos^2 \phi_{001}| = 1 \tag{7.38}$$

$$2|\cos^2 \phi_{hk0}| + |\cos^2 \phi_{001}| = 1 \tag{7.39}$$

The above equation can be transformed into Hermans orientation function and it is shown in Eqs. (7.40) and (7.41)

$$f_{100} + f_{010} + f_{001} = 0 \tag{7.40}$$

and
$$2f_{hk0} + f_{001} = 0 \tag{7.41}$$

7.5.3 Small angle x-ray scattering

Small angle x-ray scattering (SAXS) provides information about the period of the lamella stacking in semicrystalline polymers and about the layer thickness of smectic liquid crystalline polymers. The azimuthal angle dependence of the SAXS pattern provides information about the orientation of these super structures. The information about these from SAXS will be discussed in Chapter 11.

7.5. 4 Infrared spectroscopy

IR spectroscopy is very useful for the assessment of chain orientation. This spectroscopy can also analyse molecular bond directions, and can be specific to crystalline or non-crystalline regions or even fold bonds. The measurement of IR dichroism requires the use of IR radiation with parallel and perpendicular polarization to a selected reference direction. However, IR dichroism usually assumes dipole moment change corresponds to bond direction may not always be correct, only as well as bond assignments. For this analysis thin specimen is required, otherwise overlapping bonds and base line determination are difficult. The detail of IR spectroscopy is discussed in Chapter 12.

7.5.5 Raman spectroscopy

Raman spectroscopy is a scattering phenomenon, which changes the frequency of the incident light falling on a sample and the frequency difference is helpful for the structural analysis. The information available is similar in nature like that of IR spectroscopy. The molecular chain orientation as well as the orientation of different components can be analysed by means of Raman spectroscopy. However, to relate polarizability changes to directions in molecules may not be as straightforward as assuming a bond direction correspondence. For experimentation, transparent specimen with little or no fluorescence is required. The detail of Raman spectroscopy is discussed in Chapter 12.

7.5.6 Nuclear magnetic resonance (NMR) spectroscopy

NMR spectroscopy is used for determination of the structural determination of the polymers. Measurement of different proton mobility in amorphous and crystalline regions provides information regarding crystallinity of the polymer, the chain conformation or mechanism of molecular motions at different temperatures. So this information can able to distinguish crystalline and non-crystalline regions in specific cases, possibly parts of molecule determine inter nuclear vectors. However, it requires knowledge of structure to give model for positions of magnetic nuclei. The detail of NMR spectroscopy is discussed in Chapter 12.

7.5.7 Sonic velocity

When the sound pulse is sent along the length of a bundle of parallel molecules, the sound is propagated principally by the stretching of chemical bonds in the backbone of the polymer chain. For partially oriented polymer molecular chains, the molecular motion due to sound pulse transmission can have its components along and across the direction of the molecular axis. The magnitude of these two components is a function of θ between the molecular axis and the direction of sound propagation. The velocity of the sound pulse depends on the modulus, and the velocity of the sound is greater when it propagates along than transverse to the chain axis. In a semicrystalline polymer, both the crystals and the amorphous phase contribute in proportion to the relative contents. In that sense, the sonic modulus is similar to birefringence. The relationship between sonic modulus (E_s) and orientation factor for a semicrystalline polymer is given in Eq. (7.42)

$$\frac{1}{E_s} = \frac{x_c}{E_{t,c}{}^o}(1-\cos^2\theta_c) + \frac{1-x_c}{E_{t,a}{}^o}(1-\cos^2\theta_a) \tag{7.42}$$

where x_c, the degree of crystallinity, θ, angle between the molecular axis and the direction of sound propagation, and the subscript c stands for the crystalline region and a for amorphous region. $E_{t,c}{}^o$ and $E_{t,a}{}^o$ are the modulus for the propagation of the sound perpendicular to the chain in the crystalline and the amorphous fractions, respectively. For an unoriented sample, $\cos^2\theta_a = 1/3$ and Hermans oriented factor is related to $\cos^2\theta_c$ as per Eq. (7.19). With this modification, Eq. (7.42) can be modified and it is written in Eq. (7.43)

$$\frac{3}{2}(\Delta E)^{-1} = \frac{x_c \cdot f_c}{E_{t,c}{}^o} + \frac{(1-x_c)f_a}{E_{t,a}{}^o} \tag{7.43}$$

where $(\Delta E)^{-1} = E_s^{-1} - E_u^{-1}$, E_u, sonic modulus of the unoriented sample, f_c, Hermans orientation function for crystallite components and f_a, Hermans orientation function for amorphous components. This equation is also used to estimate the amorphous orientation function.

7.6   Orientation process

Orientation is the result of deformation. Mobile molecules are extended by the application of an external force. The oriented state is frozen by quenching of the sample or by rapid cooling or by a pressure. Orientation can be induced by any of the two methods and those are: (a) liquid state process and (b) solid state process. The liquid state process includes melt spinning and solution spinning of crystalline or liquid crystalline polymers. This is basically related to the production of the fibres. The solid state process includes drawing, cold drawing, extrusion or rolling. Drawing or cold drawing is a part of fibre formation process.

(a)   **Liquid state orientation process**: There are practically three states through which a fibrous polymer may pass during production. At melt or at solution, the molecular chains are in a completely disordered amorphous condition. The melt or the solution is strained and the molecules are extended from their equilibrium isotropic state. This mobile oriented liquid may be quenched by means of rapid cooling or by pressure with or without mass transfer. So the liquid process is dependent upon the oriented state being frozen with minimum relaxation. Along with orientation, the lateral and axial orders, when developed, form crystalline and amorphous configurations. In this case, all configurations have same degree of axial and lateral orders. This consists of orientation of the crystalline regions accompanying the orientation of the amorphous region. High molecular mass polymers show the longest relaxation times and are the most likely to be crystallized while being oriented and hence form extended chain crystals. Crystallization of these extended chain crystals after post formation operation results in epitaxial folded chain lamellae.

(b) **Solid state orientation process**: The solid state process involves plastic deformation of an isotropic or weakly anisotropic solid. The process includes drawing, cold drawing, extrusion or rolling. However, drawing and cold drawing are more important in fibre formation and development process for the required quality of the end product. The polymers which can be oriented are semicrystalline with deformable amorphous ($T > T_g$) and crystalline phases. Cold drawing leads to necking, i.e. the formation of a localized zone in which the unoriented structure is transformed into a fibrous structure. The neck zone travels through the specimen until the entire sample is drawn to a fibrous structure. The unoriented structure is mostly a spherulitic structure and it transforms into a fibrillar structure. The initial structure twists and breaks up into smaller crystallites which are pulled into long and thin microfibrils. The microfibrils connected with many taut tie chains. The transformation into the fibrillar structure is also accomplished by affine deformation of the molecules.

The molecular orientation of the amorphous fraction of fibres is widely accepted as a decisive factor determining many important physical and physicochemical properties of fibres. The amorphous orientation strongly affects (i) the mechanical response to the imposed force, (ii) transport process and (iii) diffusion kinetics of different chemical agents. Because of this, the examination of amorphous orientation is more important and it can be studied by the combination of x-ray diffraction and optical methods. IR spectroscopy can also directly measure the amorphous orientation.

7.7 Importance of orientation

The performance of any material depends upon (a) the mechanical properties under end use conditions, (b) the mechanical properties under fabrication conditions and (c) influence of the fabrication variables upon the structure and the fabricated material. The mechanical property of any fibre depends upon the molecular structure and the spatial arrangement of the molecules. The arrangement of the molecules includes molecular orientation. Molecular orientation occurs invariably in most mechanical fabrication operations. Fibres and monofilaments are subjected to service mainly to tensile and bending loads and consequently uniaxial orientation can provide satisfactory mechanical performance.

Molecular orientation is a process in which the polymer or the fibre is heater to some temperature above glass transition temperature, stretched at a controlled rate and cooled under tension to retain the orientation. Molecules are shifted from a preferred random coil entanglement to a relative alignment to the principal axis of stretch. Controlled molecular orientation is an asset since most physical properties are greatly improved by it. Molecular orientation simply alters the mechanical properties of polymer but also resistance to combined action of mechanical stress and environmental agents.

1. When a polymer or fibre is stressed unidirectionally while being formed, it will be oriented unidirectionally. It will be strong in the direction of orientation but weak in transverse direction.

2. Molecular orientation occurs naturally during most mechanical fabrication processes. The pattern of molecular orientation in part governed by the process kinematics – the pattern of flow and/or viscoelastic deformation which the polymer flows.

3. The effect of molecular orientation can either be favourable or unfavourable depending on direction of orientation relative to the stress encountered in service.

4. Uncontrolled orientation is a source of weakness and failure. Very seldom it will have beneficial effects.

5. Planned, controlled molecular orientation can be a valuable aid in achieving optimum properties and performance.

7.8   Deformation behaviour of fibres

Macroscopic strain applied along the fibres causes strain in the polymer chains inducing changes in bond angles and bond lengths and corresponding changes in vibrational spectra. The influence of stress on deformation of highly oriented and high performance fibres and polymers like poly(p-phenylene terephthalamide) (aramids), poly(p-phenylene benzobis thiazole), ultra high molecular weight polyethylene (UHMW-PE), carbon fibres, etc. have been analysed by studying the Raman spectra on a molecular level.

Effect of uniaxial stressing on UHMW-PE indicates that the stress and strain are macroscopic. However, the stress induces microscopic strain in the carbon backbone. For low stresses, the strain is linear. The taut tie molecules are the interfibrillar material that becomes stressed as fibrils slide past each other during creep. The macroscopic strain is transferred more effectively into C–C bonds in case of fibres with higher Young's modulus. A broader strain distribution results in relatively high stresses in some of the C–C bonds. Drawing of the fibre will induce C–C bond rupture at smaller elongation. Fibre exhibiting a narrow strain distribution will have rupture at higher elongation.

7.9   Orientation and anisotropic properties

Consider a rectangular Cartesian system of co-ordinate axis XYZ. The orientation of the C-axis of the triclinic unit cell can be described by two angles α and β (One is between C and Z axes and another projection of C in XY plane and X axis). Let γ is the projection of A axis in this plane (XY) normal to the C axis. The three angles α, β and γ are called Euler angles and define three successive rotations. The orientation can be defined in terms of α, β and γ. The orientation distribution function for all the unit cells in hypothetical crystalline aggregate will take the form $f(\alpha, \beta, \gamma)$, where

$$\int_0^{2\pi}\int_0^{2\pi}\int_0^{2\pi} f[\alpha,\beta,\gamma], \sin\alpha, d\alpha\, d\beta\, d\gamma = 1$$

Comparison of the breaking strength (BS) with the birefringence of the filament at the breaking point, it will be seen that 'The breaking strength of a filament is governed within fairly narrow limits by its birefringence i.e., by its average angle of orientation at the breaking point'. This holds well so long as the breaking strength is not unduly high. There is a very wide scattering with the higher values. The BS can approximately be related to the optical orientation factor (f) as:

$$BS\,(1-f) = C$$

where C is constant. This value depends on the fibres and is approximately equals to 45 for cellulosic fibres. The equation can be written as

$$BS = \frac{2}{3} \cdot \frac{C}{\sin^2\theta_m}$$

where θ_m is the average angle of orientation. This means that BS is inversely proportional to the average sine angle of orientation at the breaking point.

Further readings

1. M. Lewin and E. M. Pearce, *Handbook of Fibre Science & Technology*, Marcel Dekker, New York, 1985.

2. S. Eichhorn, J.W. S. Hearle, M. Jaffe, T. Kikutani (Eds.), *Handbook of Textile Fibre Structure*, edited by, Woodhead Publishing Series in Textiles, 2009.

3. I.M. Ward, *Structure and Properties of Oriented Polymers*, Springer, Netherlands, 1997.

4. I.M. Ward, *Mechanical Properties of Solid Polymers*, Wiley, Chichester, 1983.

5. A. Ziabicki, *Fundamentals of Fibre Formation: The Science of Fibre Spinning and Drawing*, Wiley, 1976.

Measurement of structures

8.1 Introduction

Technical developments in man-made fibre production, processing as well as its applications are mostly influenced by structure and structural developments of both new and existing polymers for fibre formation. The end-use properties and thus applications are dependent upon the structure developed. It is essential to know the structure of the material as well as the structure developed during processing such that the end-use properties can be controlled or modified for the specified applications. So the analysis of the structure, as well as the structural measurements is essential for understanding the relationship between the structure and properties of the material.

Fibres as well as the fibre forming polymers are typically complex morphological systems and it is composed of different characteristics like molecular weight, tacticity, crystallinity, orientation and sometimes chemical composition. Additionally, the complication of enormous macromolecular chains means that simplified descriptions are often needed to characterize polymeric materials. A detailed understanding of many of the analytic descriptions of polymeric materials is often precluded by the complexity of the situation. The specific analytic feature of a processing operation such as fibre formation of a polymeric material such as polyamide or polyester is related with the structural level like the orientation of chains or crystallites (lamellae), orientation of spherulites and the amorphous components.

A complete understanding of the fibre structure requires investigations on its structures like configuration, conformations, crystal structure and dimensions of crystallites, their arrangement, crystallite and crystallinity relationship and the nature of non-crystalline regions. This information can be obtained from various techniques like X-ray diffraction and scattering, Thermal methods, Microscopes, Small-angle Light Scattering, IR Spectroscopy, NMR Spectroscopy, Small-angle Neutron Scattering, Raman Spectroscopy, Dielectric Loss and Dynamic Mechanical Spectroscopy. Table 8.1 highlights different structures present in a fibre and the range of different experimental techniques used for investigation. The analytic description of a complex material like semicrystalline polymer or fibre is strongly dependent on the size scale on which an observation is made. The understanding of overall structure can be done by integration of information derived from a variety of sources. The molecular structure can be further interpreted.

Table 8.1   Structure and methods used for determination of structure

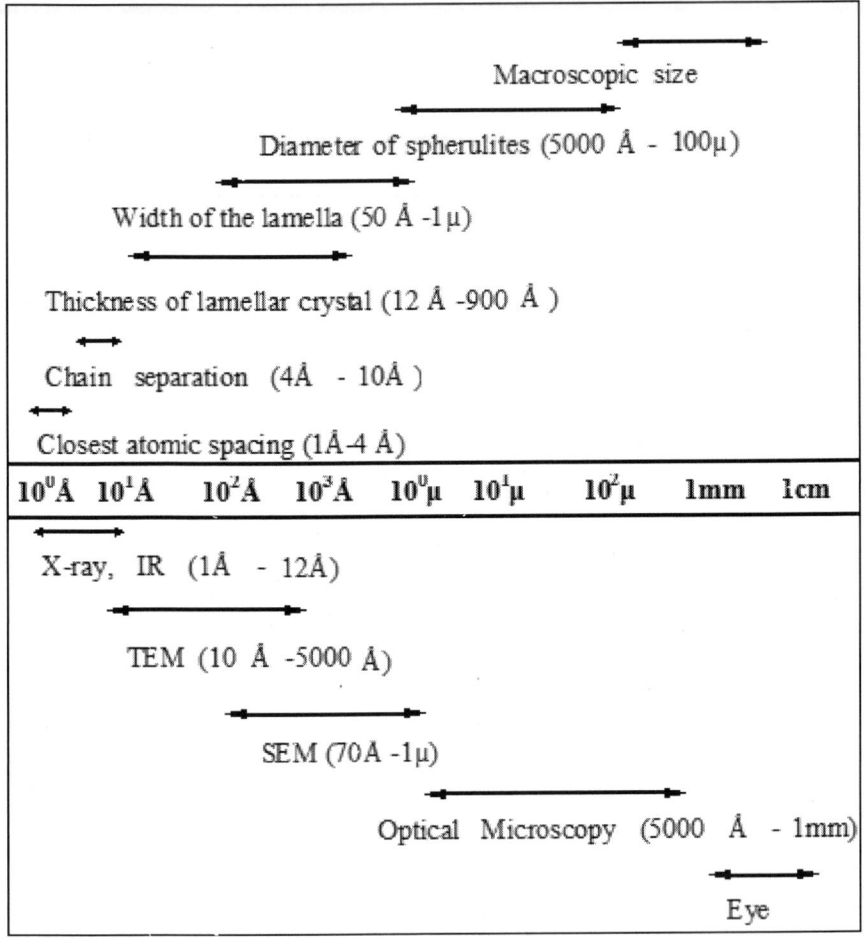

8.2   Methods for investigation of configuration and conformation

The polymeric materials like fibres are composed of chemical units. The chemical structure or the structure of the chemical units combined with the topological arrangement of the repeating units (tacticity) in a polymer chain give rise to different configuration and conformation of polymer chains. This conformation and the weak chemical associations give rise to some mechanical and vibrational features and these features can be observed spectroscopically. Generally, the coiling of monomers in a chain is evidenced by colloidal scale structure, chain persistence (local linearity) and enhancement of the ability of long chain polymers to crystallize. The chemical units give rise to spectroscopic absorption patterns which are mostly similar with their lower molecular weight counterparts. Absorption is a quantized inelastic phenomenon involving the transfer of energy from EM radiation to a material. The major techniques for the determination of chemical composition or the chemical units and molecular topology involve the absorption of electromagnetic radiation by polymers.

Spectroscopic techniques commonly used for the chemical analysis of polymeric systems. The major spectroscopic techniques used are Infrared (IR) Spectroscopy, Raman Spectroscopy, Nuclear Magnetic Resonance (NMR) Spectroscopy and to some extent Ultraviolet/visible (UV/Vis) Spectroscopy. Investigation by molecular spectroscopy methods like Infrared Spectroscopy and Raman Spectroscopy is helpful to determine the stereoregularity of the polymers. Raman Spectroscopy is more connected with physical structures like conformations and tacticity. Investigation by Nuclear Magnetic Resonance (NMR) is related to the measurement of different proton mobility. This investigation yields information on the constitution and conformation of chain molecules and molecular motion in polymers. The information related to the internal molecular motion can be derived from Dynamic Mechanical Spectroscopy, dielectric loss and neutron scattering.

IR spectrum is capable of giving indirect, but very valuable qualitative and quantitative structural information on polymer to evaluate the structural order and quantitative composition of the polymers, the analysis and identification of chemical structures, their changes, conformation, tacticity and crystallinity. The vibrational motion of the bonds may provide information on chain folding, molecular arrangement in the amorphous region like trans and gauche (cis) conformation, coiling, tie chain length distribution in the amorphous region. Raman Spectroscopy has immense potentiality. The rapid growth in the use of this technique for fibre characterization is due to the emergence of new instrumentation. Raman Spectroscopy is more connected with physical structures like analysis of conformation, tacticity, crystallinity, state of the amorphous phase, orientation, deformation. NMR spectroscopy is a very sensitive tool to study the conformations of polymer chain, as it is very sensitive to small changes in conformation. NMR spectroscopy can examine the validity of the models used for the conformational analysis and also can predict the exact conformation present in the structure.

X-ray crystallographic data derived from the crystallization of single polymer crystals indicates that polymer chains readily form regular folds. X-ray crystallographic examination supplemented by other appropriate techniques like IR spectroscopy, optical activity, NMR spectroscopy and birefringence studies can help to find the configuration and geometrical arrangement of the chains.

The configuration of the macromolecule in dilute solutions can be evaluated from light scattering or small-angle x-ray scattering. In case of bulk amorphous polymers, similar observations can be available from small-angle neutron scattering. The anisotropy or density fluctuations can also be available from light scattering or small-angle x-ray scattering. For polymer chains in bulk amorphous state, the environment of the chain is composed of other chains of the same compositions. So the configuration in bulk is possible by means of small-angle neutron scattering. This is the only technique available so far for investigating in the bulk by means of a straightforward structure method.

8.3   Methods for investigation of crystallites and crystallinity

In the polymeric materials, local chain structure is sufficiently regular to give rise to a crystalline phase. Entanglement of chains, chain branching and the presence of end groups prevents complete crystallization of a polymer. Polymeric materials which display crystallinity are always described by a multi-phase model, i.e. semicrystalline, which includes an amorphous and crystalline phase in coexistence. Low transport coefficients and chain folding give rise to nanoscale crystallites which are best observed by Transmission Electron Microscope (TEM), Small-angle X-ray Scattering (SAXS) and also by Raman spectroscopy (Longitudinal Acoustic Modes, LAM). Fibrillar crystallites in polymers lead to colloidal to optical scale structures, spherulites, which are generally centrosymmetric and which display radially oriented birefringence. These micron scale structures are best observed using optical microscopy, Scanning Electron Microscope (SEM), Small-angle Light Scattering (SALS). Elastic interaction between EM radiation and a material is possible and this gives rise to diffraction and scattering phenomena. The small crystallite size and dominance of crystalline orientation lead to several unique analytic approaches in the analysis of X-ray Diffraction (XRD) data in these materials.

Thermal methods are simple, fast, efficient and more informative and so it is used even for routine quality control. Melting endotherm, supported by crystallization behaviour can be analysed in terms of crystallite distribution, crystallite characteristics and its stability as well as perfection. XRD method is the most important method to detect the types of crystals and crystallinity. Crystallinity from density is a simple, non-destructive and economical method. It is only method to measure the variations in the crystallinity thereby variations in the process. Crystallinity measurement from IR spectroscopy offers the advantages for a complete understanding of the fibre structure in terms of crystal structure, chain folding, arrangement of crystallite and the nature of non-crystalline regions.

SAXS is a critical technique for the description of polymeric materials since diffraction at small angles is associated with the colloidal to nanoscales which is the size range of a typical polymer chain. The colloidal scale is also associated with polymer crystallites (lamellae) and microphase separated block copolymer structures. SAXS yields valuable information on the size, shape and arrangements of large particles having characteristic dimensions in the order of hundreds and thousands of angstrom units. The majority of the SAXS studies of polymers and fibres are on the two-phase microstructure like shape and size of crystallites, voids and intercrystalline spaces. SALS and polarizing microscope can be used to determine the shape and size of the superstructure.

Intermediate size structure can be seen by electron microscope but the technique is limited to the inspection of surface or thin sections. Information about structure up to 400 Å in size may be obtained by SAXS if there is a periodicity in density. SALS provides information about the structures having size of the order of wavelength of light, in the region of 1000 Å to 10,000 Å. It also provides information about fluctuation in density of scattering material but in size range much larger than that can be studied by x-ray. X-ray involves the inner electrons of atom while light scattering involves the outer valence electrons.

The crystalline regions vary in size and perfection. Information on the size may be obtained from XRD since the breadths of the lines in the pattern depend on crystal size. Crystallite sizes can also be obtained by scattering of x-rays at low angles, i.e. SAXS and also by TEM. SAXS is generally used to study the shape and dimensions of crystallites. TEM provides the structural information relating to crystallites, crystalline phase structure and dispersed phase structure. Electron diffraction is also used to study the structure of crystalline and amorphous state with small sample size and within short exposure time. This is generally carried out in a TEM. The electron microscope can be used to observe individual macromolecules, to study the supermolecular structure and to see polymer single crystals. It is with the aid of electron microscopy that spherulites, fibrils and other elements of the supermolecular structure have been studied in detail. Even SALS with wavelengths in the region of 5000 Å can provide information on supermolecular structure.

Crystallinity is best regarded as one of the molecular based properties. The higher levels of mechanical properties and the better retention of useful properties with increasing temperature are attained due to crystalline character of the fibres. Crystallites are regarded as compact reinforcing members that are less readily deformed and less readily permeated by foreign molecules because of their specific volume.

8.4   Methods for investigation of amorphous state

Small-angle Neutron Scattering (SANS) is very helpful in the study of amorphous state of the fibres. SANS gives evidence for the disorderedness and yields information on dimensions in the range of 30–300 Å. The application of SANS is related to the characteristics or changes in radius gyration. The radius of gyration of the deutrated polymers dispersed in the unlabelled host invariably turns out to be in close agreement with value for the isolated, unperturbed chain, as determined in a θ solvent. Differences between different conformation models are most pronounced in the short-range order region covered by Intermediate Angle Neutron Scattering (IANS). Even, SAXS and electron diffraction can confirm the presence of nodules or bundles in amorphous region.

The structure of amorphous polymer can be investigated by application of various experimental techniques like light scattering, SANS, SAXS and XRD. Crystallinity can be derived from density, heat of fusion, XRD, NMR, IR spectrum. The last three methods are based on a common principle that the measured diagram or spectrum is resolved into the different contributions from the crystalline and disordered regions of the sample. The mass fractions in the different states of order follow from the weights of the respective components, either directly or after calibration.

8.5   Methods for investigation of orientation state

The state of orientation of each individual molecular chain or the structural units can be measured by XRD methods. Birefringence of a fibre is also related to how chains are oriented and regarded as a parameter of the average orientation of all the chains whether in the crystalline or amorphous region. These parameters are extremely important to correlate and predict the mechanical properties with structure. Sonic modulus and/or sonic velocity help us to measure orientation of the molecular chains like that of the birefringence.

8.6   Overall observations

All analytic techniques are based on specific physical principles which serve as a guide to understanding the basic limitations of the techniques. Multiple techniques are also available to describe the same property of a polymeric material and it is essential to understand the physical basis of characterization techniques for the relationship between structure and chemical composition and properties. There are several analytic techniques available to describe orientation. These include construction of pole figures from XRD scans, calculation of orientation functions from XRD data, calculation of orientation functions from IR, NMR or Raman data, and measurement of the optical birefringence using polarized light. Each of these techniques will yield a different value for the orientation function. Similarly, even the value for the degree of crystallinity will depend on the technique which is used, i.e. XRD, differential scanning calorimetry, density or IR. This makes a firm understanding of the physical basis of analytic techniques critical to their application in polymeric systems.

All properties of polymeric systems display dispersion due to (1) the limited ability of the system to produce monodisperse and perfect chemical structure as well as (2) the dominance of kinetics in processing of high molecular weight materials. Dispersion of structure and chemical composition means that the best tools to describe polymeric materials are always statistical in nature. For example, a low molecular weight organic has a specific melting point, while a polymer displays a range of melting with an onset, a peak and a maximum melting temperature. Such a melting spectrum might best be described by a Gaussian function with a standard deviation and mean.

Statistical analysis or statistically distributed systems include the major distribution functions and mathematical descriptions of the propagation of error in data sets. Analytic descriptions of polymers are of no use unless some description of the expected error associated with the analytic results is presented. Additionally, it is expected that any analytic determination of these materials will be subject to a large range of statistical variation between samples as well as an innate distribution associated with the polydispersity of the material of itself.

Further readings

1.   D. Campbell, R. A. Pethrick, J. R. White, *Polymer Characterization Physical Techniques*, Chapman and Hall, 1989.

2.   *Encyclopedia of Polymer Science and Engineering*, Wiley, New York, 1986.

3. M. Lewin and E. M. Pearce, *Handbook of Fibre Science & Technology*, Marcel Dekker, New York, 1985.

4. H. Tadokoro, *Structure of Crystalline Polymers*, Wiley, New York, 1979.

5. Linda C. Sawyer, David T. Grubb, Gregory F. Meyers, *Polymer Microscopy*, Springer, 2008.

6. Oliver H. Seeck and Bridget Murphy, *X-Ray Diffraction: Modern Experimental Techniques*, CRC Press, 2015.

7. L. R. Alexander, *X-Ray Diffraction Methods for Polymer Science*, Kreiger Publication, 1979.

8. R.S. Stein, *Newer Methods of Polymer Characterization*, R. Ke (Ed.), Interscience Publishers, New York, 1964, p. 185.

9. A. Giunier, G. Fournet, C. Walker, and K. Yudowich, *Small-Angle Scattering of X-ray*, Wiley, New York, 1955.

10. J. L. Koenig, *Spectroscopy of Polymers*, Elsevier, 1999.

Microscopy methods

9.1 Introduction

Microscopy is the study of fine structure and morphology of objects with the use of a microscope. The successful microscope analysis depends upon the use of specialized sample preparation and observation techniques. There are basically two types of microscopes, i.e. (1) Optical or Light microscopes (LM) and (2) Electron microscopes (EM), depending upon whether light rays of electrons are used. Further, electron microscopes are mainly of two types, i.e. (a) Transmission Electron Microscope (TEM) and (2) Scanning Electron Microscope (SEM). Different microscopes available are: (a) Light microscope (LM) or Optical microscope (OM), (b) TEM and (c) SEM. The basic properties of these three different microscopes are given in Table 9.1.

Table 9.1 Properties of LM, SEM and TEM

	LM	SEM	TEM
Radiation	Visible light	Beam of electrons	Beam of electrons
Wave length of radiation	400–700 nm	<1 nm	< 1 nm
Resolution (nm)	300	10	0.2
Focus of radiation	Lenses made of glass	Electromagnets	Electromagnets
Image formed by	Light	Selective absorption of electrons	
Magnification range	2–2000	$20–10^5$	$200–2 \times 10^6$
Working range	1 μm–1 mm	10 nm–1 mm	1 nm–10 μm
Can observe	Surface	Surface	Bulk

Specific environment	Ambient	High vacuum	High vacuum
Specimen preparation	Easy	Easy	Difficult

Modern electron and optical microscopy is used to study materials where the problem is invisible to the naked eye. The choice of which microscopy technique is used is dependent on the task to be performed and the information required from the samples. In microscopy, resolution is the key parameter, which indicates its working range and the information available from the instrument. Resolution is the minimum distance between two object features at which they can still be seen as two features.

9.1.1 Applications

The microscope is used for various applications and it depends upon the facilities of the microscope. Objects of different sizes can be assessed as per the facilities of the microscope. The objects can be anything from the polymer molecules to the woven structure of the textile materials. The different objects assessed by different microscopes are mentioned in Table 9.2.

Table 9.2 Object size assessed by microscopes

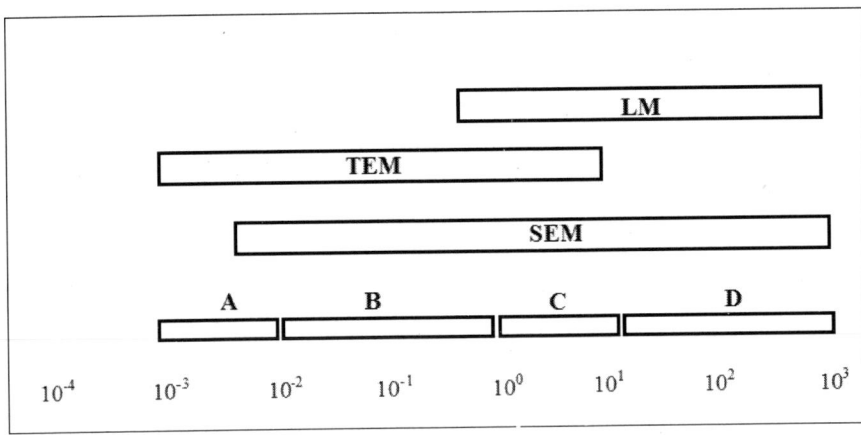

Object size (μm)

Zone A: Polymer molecule, coil, domain, crystalline and amorphous domain

Zone B: Fibrils, fillers, pigments, nucleating agents

Zone C: Spherulites, fractured surface, agglomeration of particles

Zone D: Pores, fibres, woven structures

9.2 Light microscopy

A light microscope magnifies the object by means of an objective lens and an eyepiece. The object lies slightly outside the focal length of the objective lens. When viewed through the eyepiece, a virtual, magnified inverted image can be visualized. The simple instrument in an optical microscopy is compound microscope where the object is magnified to get a magnified image. The degree of magnification depends upon the ratio of the size of the image of the object seen with the instrument to that seen with the unaided eye. The magnification can be done by means of the objective lens and the eyepiece. The objective lens created primary magnification and the total magnification is created by both objective lens and eyepiece. The primary magnification (M_1) by the objective is shown in Eq. (9.1). Total magnification (M) is generally equal to the visual length of the image seen with the instrument divided by the visual angle of the object seen directly. The expression for total magnification is given in Eq. (9.2)

$$M_1 = \frac{g}{f_o} \qquad (9.1)$$

$$M_1 = \frac{g \cdot D_v}{f_o \cdot f_e} \qquad (9.2)$$

where g is the optical tube length; f_o, focal length of the objective; f_e is the focal length of the eyepiece and D_v, least distance, i.e. 250 mm.

9.2.1 Parameters of light microscope

The optical performance of microscope is generally characterized by the (a) Resolving power (R), (b) Useful magnification (M) and (c) Depth of focus (T).

(a) Resolving power

The diffraction of light waves through a slit is expressed by Eq. (9.3).

$$\sin \alpha = \lambda / R \qquad (9.3)$$

where λ is the wavelength of light and R, the resolving power smallest resolvable separation of two point objects. As per Eq. (9.3), the resolving power of a light microscope is shown in Eq. (9.4a) in vacuum and in Eq. (9.4b) in a medium having refractive index.

(1) In vacuum or air, $\qquad R = \lambda/\sin \alpha \qquad (9.4a)$

(2) In a medium of refractive index η, $R = \lambda/\eta \cdot \sin \alpha \qquad (9.4b)$

The numerical aperture is $\eta \cdot \sin \alpha$. If $\alpha = 90°$, $\sin \alpha = 1$ then $R = \lambda / \eta$. The refractive index of immersion oil is 1.4 and λ for green light is 0.55 µm. So the resolving power of the light microscope with oil immersion is $0.55 / 1.4 = 0.4$ µm.

(b) Useful magnification

Useful magnification refers to the limiting magnification beyond which no additional resolving power can be achieved. A magnification in excess of useful magnification is known as empty magnification. This indicates that the resolving power of the human eye is involved and this depends on the range of distinct vision (250 mm) and the viewing angle of the eye ($>1'$) for a size up to 100 µm. The useful magnification (M) is usually expressed as the ratio of the resolving power of the human eye (R_e) with respect to the resolving power of the microscope (R_m) [Eq. (9.5)]

$$M = \frac{R_e}{R_m} \qquad (9.5)$$

Resolving power of human eye is approximately 100 µm. If the resolving power of the microscope is 0.4 µm, then $M = 250$.

(c) Depth of focus

The depth of focus (T) of a microscope depends on the beam aperture α and the resolving power R.

$$\tan \alpha = \frac{R/2}{T/2} = R/T \qquad (9.6)$$

or $$T = R/\alpha \qquad (9.7)$$

If $R = \lambda / \eta \cdot \sin \alpha$, η is approximately 1.0 and at small angle, $\sin \alpha \to \alpha$,

Then $$T = R \cdot [(2/\lambda^2) - 1] \qquad (9.8)$$

9.2.2 Types of microscopes

There are different types of light microscopes for different applications. These may be transmission, polarized, phase contrast, fluorescence or interference microscopy. The important types of the microscopes with their applications are discussed below.

9.3 Transmission light microscopy

In transmitted microscopy, a collimated light beam passes directly through the sample and into the objective lens of the microscope. The magnified image may be further enlarged or inverted by intermediately optics. The image may be viewed through microscope eyepiece (compound microscope) or projected on a screen (projection microscope). Morphological features which scatter light or give different optical density are visible. In light microscopy, the resolution is limited by the wavelength and the quality of the light and is approximately 0.2 µm. So the specimen must be sufficiently thin that an appreciable fraction of light is transmitted and that internal structures are not excessively overloaded. For opaque or thick specimen, reflected light microscopy is generally used to examine variations in surface structure. Light microscope facilities provide low magnification (8–1000 X) of fibres and textiles that can be used to examine fibre surface characteristics, scale pattern, debris and wear; identify individual fibre types and measure their shape and dimensions. In addition, LM is useful for the determination of the construction and function of fibrous assemblies, diagnosis of chemical and microbial damage, and evaluation of processing defects and mechanical degradation. The principle of transmission light microcopy is shown in Fig. 9.1.

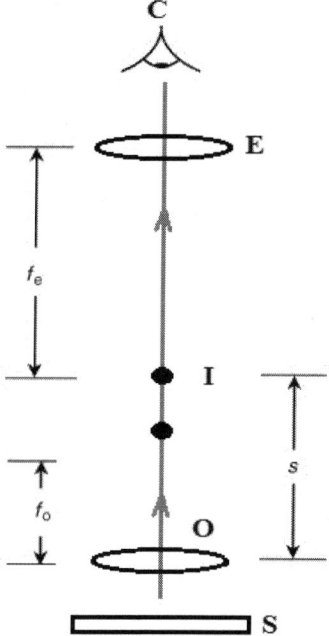

Figure 9.1   Transmission light microscopy: (C) Observer, (E) Eyepiece, (I) Real image from objective, (O) Objective, (S) Sample for observation.

9.4   Reflected light microscopy

In incident light microscopy the light is reflected off the sample. When specimens are opaque or excessively thick, reflected light microscopy may be used to examine variations in the surface structure. Contrast arises from the variations in surface reflectivity.

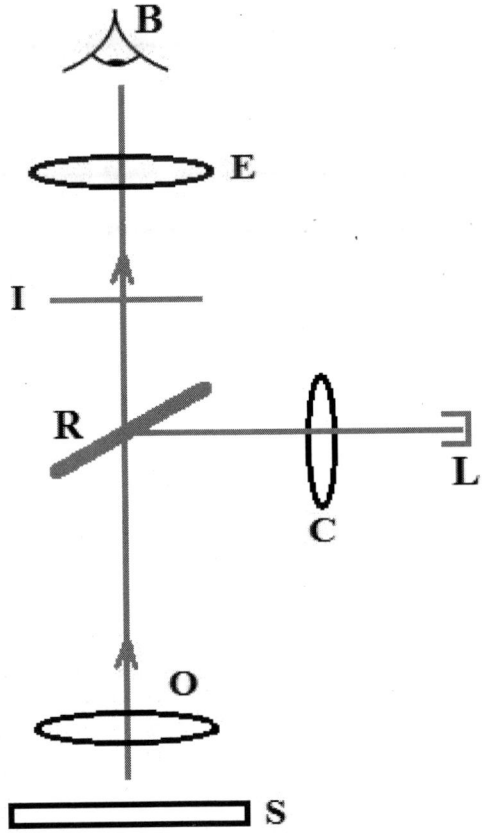

Figure 9.2   Reflected light microscopy: (B) Observer, (E) Eyepiece, (I) Real image from objective, (R) Reflector, (C) Condenser, (L) Light source, (O) Objective, (S) Sample for observation.

9.5   Dark field microscopy

In this microscopy, dark field may be carried out using either transmitted or reflected light. In contrast to bright field illumination, directly transmitted or reflected light is prevented from entering the objective. In transmission, this is accomplished by placing a circular stop above the condenser to block the central beam. In reflection, special objective apertures are used to provide very oblique illumination of the specimen. When viewed in dark field, object, which reflects or scatters light, is visible against a dark background and marked gains in contrast are often obtainable.

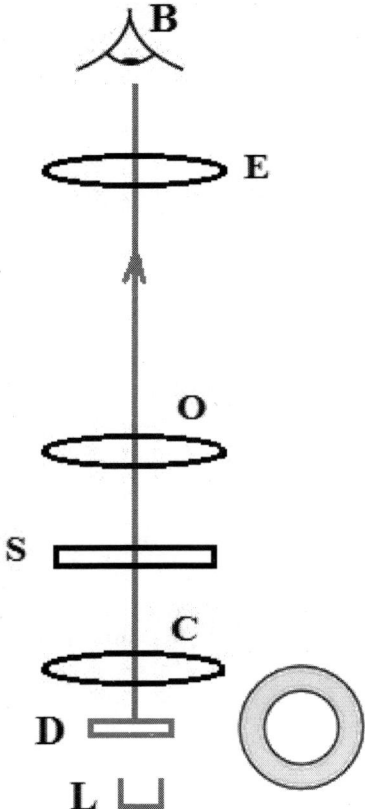

Figure 9.3 Dark filed microscopy: (B) Observer, (E) Eyepiece, (C) Condenser, (L) Light source, (O) Objective, (D) Dark field stop, (S) Sample for observation.

9.6 Fluorescence microscopy

Fluorescence microscopy is a form of light microscopy that uses fluorescent dye to label and locate specific cell components. In this microscopy, fluorescent dyes bind to particular chemicals and are used in conjunction with specific light sources and filters. The sample is brought into contact with a fluorescent liquid like rhodamine. The liquid penetrates in the cracks and pores and thus permits these regions to fluorescence when illuminated by UV light in microscope. It is then possible to image the sample, showing where particular chemicals are located. It can provide valuable applied research information in areas such as dyeing, but also can be used in fundamental research for the location of cell types and to map protein expression. In practice, a filter corresponding to the wavelength of the fluorescent radiation is placed in the path of the rays to absorb or reflect the exciting rays.

9.7 Confocal microscopy

Confocal microscopy is an optical imaging technique for increasing optical resolution and contrast of a micrograph by means of adding a spatial pinhole placed at the confocal plane of the lens to eliminate out-of-focus light. It aims to overcome the limitations of traditional fluorescence microscopy. A confocal microscope uses point illumination and a pinhole in an optically conjugate plane in front of the detector to eliminate out-of-focus signal – the name 'confocal' stems from this configuration. It enables the reconstruction of three-dimensional structures from the obtained images. The confocal laser scanning microscope (CLSM) allows a fibre to be optically sectioned, giving greater flexibility and control than standard fluorescence microscopy, with significantly improved image clarity and resolution.

9.8 Polarized microscopy

Light is a transverse wave motion. The light particles vibrate up and down and at right angles to the direction of propagation of the wave. The natural unpolarized light can be looked upon as a mixture of waves linearly polarized in all possible transverse direction with random orientation. Such a light is called 'symmetrical light'. When a beam of symmetrical light is incident on a polarizer, cut parallel to its optic axis, the crystal (polarizer) allows only those to pass through it which are parallel to its optic axis. The light emerging out of the polarizer has its vibrations confined to one plane and the incident light loses its symmetry. It is known as plane polarized light. The plane polarized light cannot get through a second polarizer if the optic axis is at right angles to that of the first polarizer. So the second polarizer can detect polarization as well as the direction of vibration in the polarized light. This polarizer is known as 'analyser'. Any device which can polarize light can also be used as analyser. The intensity of the light passed by a system of polarizer and analyser arranged in optical series should vary from zero to a maximum as the polarization planes of polarizer and analyser varies from perpendicular to parallel. The arrangements are commonly known as crossed polars and open polars. Polaroid is the trade name given to polarizing sheets or plates, e.g. tourmaline crystal, marks plates, etc.

In polarized light microscopy, the illuminating source is plane polarized before it impinges on the sample. A second polarizer (analyser) is inserted in the reflected or transmitted beam at 90° to the first polarizer. Only those materials which cause a partially depolarization of the light are visible and thus it can be observed. Polarizing microscope is used for the spherulitic structure of the crystalline polymer and also for birefringence of the material. The sample is generally placed with 45° and 135° of its major refractive indices to the direction of polarizer. The incident, i.e. plane polarized light is split into two polarized waves of equal amplitude parallel to the direction of the major refractive indices. These two waves then traverse the sample with different velocities. Upon emerging the sample, they combine to form an elliptically polarized wave, only the portion polarized parallel to the analyser is transmitted. The principle of polarized microscope is mentioned in Fig. 9.4.

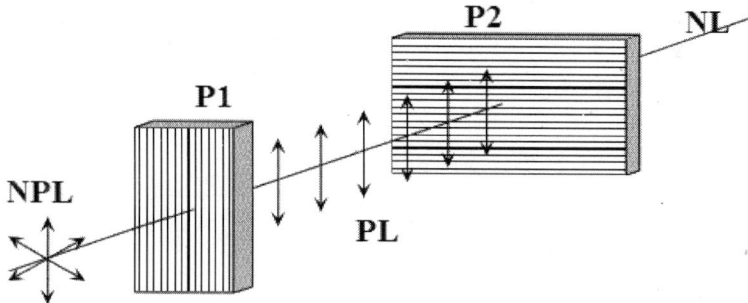

Figure 9.4   Principle of polarization microscopy: (NPL) Non-polarized light, (PL) Polarized light, (NL) No light, (P1) Polarizer, (P2) Analyser (Polarizer).

Birefringence (Δn) is a measure of the total molecular orientation of a system. It is the difference in the principal refractive index ($n\|$) and ($n\perp$) to the stretch directions for uniaxial oriented specimen. The refractive index is a measure of the velocity of the light in the medium and is related to the polarizability of the chains. The difference in the optical path lengths $\Delta n\cdot d$ in the sample of thickness d is denoted by the path difference (ΔT) ($\Delta T = \Delta n\cdot d$). The path difference can be measured directly by compensation method, which will help to measure the value of birefringence. The detail of measurement of orientation from the birefringence was discussed in Chapter 7.

9.9   Phase contrast microscopy

Phase contrast microscopy has been most effectively applied in transmission. This technique depends upon the fact that the differences in the refractive indices of two transparent phases produce small phase shifts in the light exiting from each component. The microscope contains two optical devices: The Phase Annulus in the condenser and the Phase Plate in the objective. By the use of an annular ring condenser and a phase retarding absorption ring (placed above the objective), these phase differences can be converted to intensity variations in the observed image.

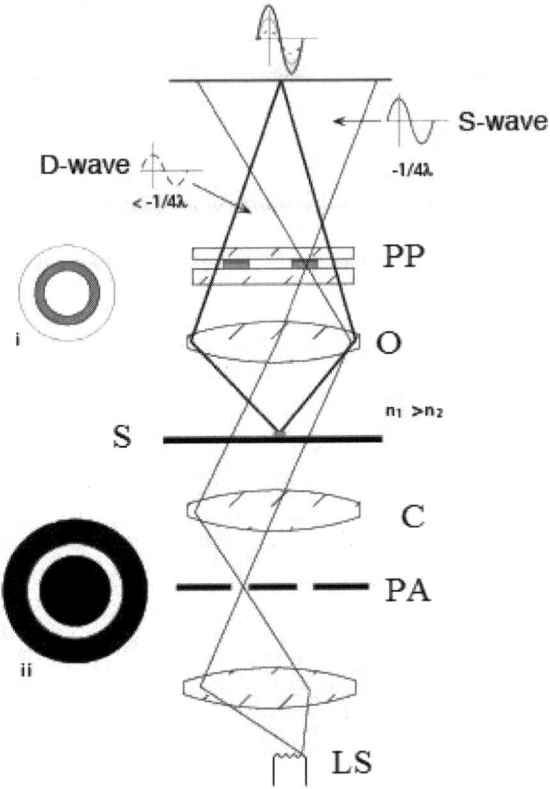

Figure 9.5   Principle of phase contrast microscopy: LS: Light source; C: Condenser lens; O: Objective lens; PA: Phase Annulus; PP: Phase Plate; S: Sample.

With no sample present, the illuminating light continues as an undiffracted plane wave (the 'Surround wave' or 'S-wave') that is focused on the Phase Plate, and uniformly spreads over the image plane. The light is retarded in phase by 1/4 wavelength. The phasing sample is referred because of the phasing is due to a difference in refractive index between sample and background. With a phasing sample, refraction and/or diffraction of light produces a spherical wavefront (Diffracted wave or 'D-wave') that shifts the sample image rays away from the phasing mask on the phase plate. The D-wave (sample light) is focused on the image plane, while the S-wave fills the image plane. Optical path differences in the sample (versus background) results in sample light being retarded in phase a variable amount up to 1/4 wavelength. When combined at the sample plane, background light (exactly −1/4 wavelength) constructively interferes with sample light (<−1/4) increasing the amplitude of the wave through constructive interference. This results in bright sample image on a darker background. This type of phase contrast (bright sample on a darker background) is known as Negative Phase Contrast. Positive Phase Contrast has the inverse phase relationships between the background (S-wave) phase and sample (D-wave), producing a dark sample image on a uniform gray background. Figure 9.5 highlights the principle of phase contrast microscopy.

The intensity variations are qualitatively proportional to optical path variations in the object. Transparent two phase systems, in which the components having significantly different refractive index, can be investigated using phase contrast microscopy. Observations of objects only alter the phase relationship but not the amplitude of the light. Introduction of a plate in the image side focal plane of the objective produces ±90° phase shift in the zero-order diffraction maximum. In general, condensers and phase retarders are used to measure phase differences. For high contrast, phase plate like $\lambda/2, \lambda/3, \lambda/4$ can be used to shift the light of the wave.

9.10 Interference microscopy

The phenomenon of interference can be explained by assuming light to be a wave motion. At any point, where one or more waves cross one another, they are said to interfere. The interference can either be (a) constructive interference or (b) destructive interference.

(a) Constructive interference: Two waves reach in phase with each other and it increases the resultant amplitude with enhanced light intensity.

(b) Destructive interference: When two wave trains reach out of phase by 180°, the resultant displacement of the medium is 0 and the results in complete darkness.

According to the theory of interference, the resultant disturbance or displacement of the medium at any point or at any instant can be found by algebraically adding the instantaneous displacement that would be produced at the point by the individual wave trains if each were present alone.

In interference microscopy, monochromatic light from a single source is divided by means of a beam splitter or a birefringent plate. One of the beams passes through the sample and then recombined. So contrast results will be obtained due to interference of two beams. This interference is helpful to study detailed variations in surface topography. Also, qualitative measurements of the path length differences are possible. The principle of interference microscopy is illustrated in Fig. 9.6.

Another modification of this technique is polarizing the two beams in mutually perpendicular directions through the use of a birefringent plate, allowing them to pass through the sample and then recombining them with a second birefringent plate relax the monochromatic light source. The principle of multiple beam microscopes is that transparent materials with 80–90% reflected capacity is generally used. It will give both constructive and destructive interferences.

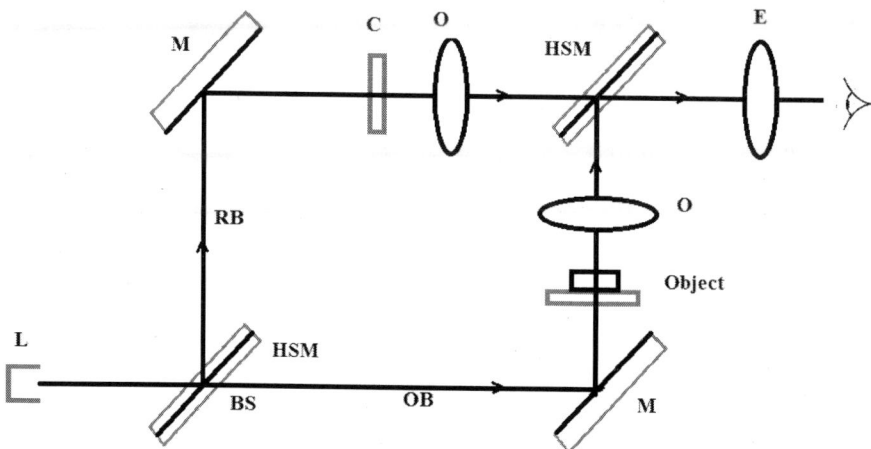

Figure 9.6   Principle of interference microscopy: L: Light source; M: Mirror; HSM: Half silvered mirror; O: Objective lens: E: Eyepiece; OB: Object beam; RB: Reference beam; C: Comparison slide.

9.11   Advantages and disadvantages of light microscopy

Optical microscopy provides a rapid and informative overview. The sample preparation is ease, quick, non-destructive and without any dimensional changes. Some quantitative measurements like thickness, refractive indices, roughness and orientation can be conducted by optical microscope only. The interpretation of the result is also simple. Important information can be obtained in short times with minimum capital expenditure and relatively ease sample preparation. This information is very vital for further investigations. However, optical microscopy techniques have limited resolution, limited magnification range and a decreasing depth of field with increasing magnification.

9.12   Electron microscopy

It is a special class of microscopes that use electrons to produce magnified images, especially of objects having dimensions smaller than the wavelengths of visible light, with linear magnification approaching or exceeding a million. Electron microscope is a device for forming greatly magnified images of objects by means of electrons. In case of electron microscopes, electrons are used instead of lights as source. Electron microscopes are used to provide high quality images at very high magnifications. The simplest and most common form of microscopy is optical or light microscopy. The highest useable magnification available with normal optical microscopes is around 1000 times. By using electrons rather than light it is possible to attain much higher magnifications up to the range of 300,000 times to 1,200,000 times. When electrons interact with the specimen in an electron microscope secondary and backscattered electrons as well as x-rays are emitted. By utilizing the various emissions valuable information can be gathered from the specimen.

9.12.1 Advantages of electron microscopes

1. They have a higher resolution and so are able of a higher magnification (up to 2 million times). Light microscopes can show a useful magnification only up to 1000–2000 times. This is a physical limit imposed by the wavelength of the light. Electron microscopes therefore allow for the visualization of structures that would normally be not visible by optical microscopy.

2. Depending on the type of electron microscope, it is possible to view the three dimensional external shape of an object (SEM).

3. In SEM, due to the nature of electrons, electron microscopes have a greater depth of field compared to light microscopes. The higher resolution may also give the human eye the subjective impression of a higher depth of field.

9.12.2 Disadvantages of electron microscopes

1. They are extremely expensive.

2. Sample preparation is more complicated and more elaborate.

3. Coating is necessary for the specimen with a very thin layer of metal.

4. The sample must be completely dry. This makes it impossible to observe living specimens.

5. It is not possible to observe moving specimens.

6. It is not possible to observe color. Electrons do not possess a color. The image is only black/white.

7. Skill is required for sample preparation and experimentation.

8. The space requirements are high.

9. Maintenance costs are high.

Electrons have both wave and particle properties and their wave-like properties mean that a beam of electrons can be made to behave like a beam of radiation. The wavelength is dependent on their energy, and so can be tuned by adjustment of accelerating fields, and can be much smaller than that of light, yet they can still interact with the sample due to their electrical charge. Electrons are generated by a process known as thermionic discharge in the same manner as the cathode in a cathode ray tube. The generation can be accelerated by an electric field and focused by electrical and magnetic fields onto the sample. The electrons can be focused onto the sample providing a resolution far better than is possible with light microscopes, and with improved depth of vision. Details of a sample can be enhanced by means of heavy metals such as osmium or lead or uranium. These metals can be used to selectively deposit heavy atoms in the sample and enhance structural detail; the dense nuclei of the heavy atoms scatter the electrons out of the optical path. The electrons that remain in the beam can be detected using a photographic film, or fluorescent screen among other technologies. So areas where electrons are scattered appear dark on the screen, or on a positive image. Like optical microscope, the electron microscope can be characterized by (a) Resolving power (R), (b) Useful magnification (M) and (c) Depth of focus (T).

(a) Resolving power

The resolving power can be derived as per the following equation:

$$eU = 1/2 \cdot mv^2$$

or
$$v = [(2e/m)U]^{1/2}$$

where e is the electron charge; m, electronic mass; v, velocity of electrons and U, accelerating voltage.

The wavelength of matter waves is given by De Broglie's equation

$$\lambda_e = (h/m \cdot v)$$

where λ_e is the wavelength of electron and h, Planck's constant

so $\lambda_e = [(h^2)/(2emU)]^{1/2}$

This is approximately equivalent to

$$\lambda_e = [1.5/U]^{1/2} \text{ in nanometre and } U \text{ in volt}$$

Wavelength achieved is actually higher due to relativistic velocity correction

For example, it can be 0.0060 nm when $U = 40$ kV and when $\lambda = 0.0025$ nm, $U = 200$ kV. The resolving Power is in the range of 0.2–0.4 nm

(b) Magnification

The magnification of the electron microscope can be calculated similarly like that of the light microscope. For calculation, Eq. (9.5) ($M = R_e/R_m$) can also be used, where R_m is treated as the resolving power of microscope, which is 0.2 nm. So M is 100 µm/0.2 nm = 500,000. For comparison, the resolving power of TEM is 0.2 nm, where as it is 10 nm and 400 nm, respectively for SEM and LM, respectively. The magnification accordingly will be 500,000; 10,000 and 250, respectively. Figure 9.7 shows the schematic diagram of EM and LM. The sample is placed directly in the beam. Collimation, focusing and magnification are carried out by means of magnetic lenses. The power of the lenses can be varied by varying the voltages.

Electrons are strongly scattered by all forms of matter, including air. Hence the entire instrument or the microscope must be evacuated to about 10^{-4} mmHg (10^{-7} atm or 10^{-2} Pa). The lenses are basically electric or magnetic fields, symmetrical about the axis of the instrument, that have the property of bending the electron paths toward the axis. Most electron microscopes employ magnetic lenses for the highest resolution and magnification. However, good results have also been obtained with electron microscopes employing unipotential electrostatic lenses and magnetic lenses excited by permanent magnets. For 50–100 kV electrons, which are commonly employed in electron microscopes, the wavelength range is 0.0053–0.0037 nm. Hence, even though a cone of radiation with an aperture angle less than 0.01 radian contributes to an image of optimum sharpness, object separations smaller than 0.3 nm have been resolved with the electron microscope. Thus the electron microscope has several hundred times the resolving power of the light microscope. Similarly, whereas the maximum useful magnification of the light microscope is about 2000, that of the electron microscope may approach 1,000,000. The maximum useful magnification is the least magnification of the image that reveals to the observer all the specimen detail that the microscope is capable of conveying. Electrons are commonly emitted from the tip of a fine tungsten-wire hairpin filament or, to further reduce the size of the effective electron source, from a sharply pointed segment of wire welded to the filament tip. The filament is maintained at a carefully stabilized negative potential of 50–100 kV with respect to the remainder of the instrument. Electrons enter the instrument through an anode aperture. The intensity and convergence of the electron beam that is falling on the specimen are adjusted by varying the coil current of the condenser lens. Image contrasts are formed by the scattering of electrons out of the narrow cone that contributes to the formation of the image; denser or thicker portions of the specimen scatter more electrons and hence appear darker in the image. The sharpness of the image observed on the screen is adjusted by varying the objective coil current, and its magnification by varying the projector coil current. Both currents must be carefully stabilized to yield high resolution. In addition to the standard transmission microscopes operating at 50–100 kV, a number of very high-voltage instruments have been constructed (for operation up to 1500 kV). The advantage of high-voltage electron microscopy does not lie in greater resolving power but in increased penetration, which is particularly valuable in the direct study of metal sections prepared with a microtome.

Electron microscopes serve primarily two purposes: the visual examination of structures too fine to be resolved with ordinary, or light, microscopes, and the study of surfaces that emit electrons. The first function made transmission electron microscopes as an essential research tools. Beginning in the 1960s the scanning electron microscope came to play an increasingly important role in the study of the surfaces of solid objects at more moderate magnifications. Based on this, there are two types of electron microscope, i.e. (a) TEM and (b) SEM

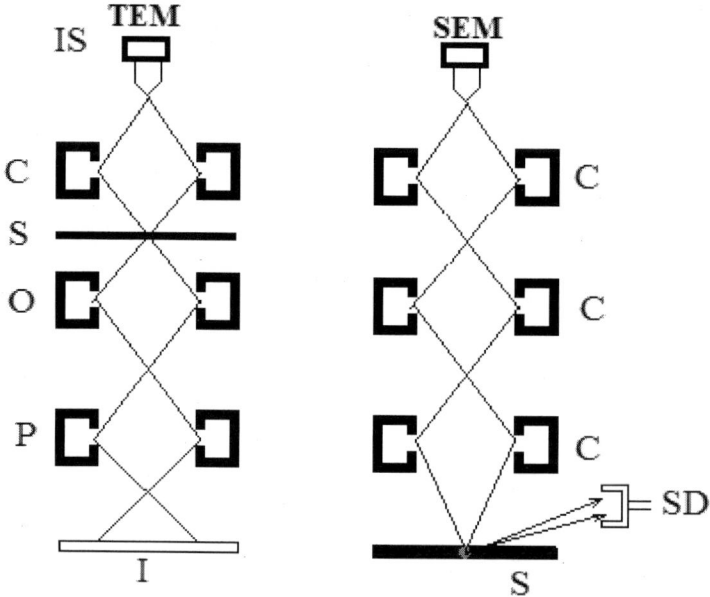

Figure 9.7   Principle of electron microscopy: TEM: Transmission Electron Microscopy; SEM: Scanning Electron Microscopy; IS: Illumination source; C: Condenser lens; S: Sample; O: Objective lens; P: Projective lens; I; Image plane; SD: Signal detector.

9.13 Transmission electron microscopy (TEM)

Transmission Electron Microscopy (TEM) is used for analysis of the structure of the materials with the help of the electrons. The electrons in a transmission electron microscope pass through the sample and are imaged on a fluorescent screen at the bottom of the microscope column. The sample has to be thin enough to allow the transmission of electrons. In order to slice the sample thinly, an ultramicrotome is used. In some instances it may be necessary to enhance the contrast in a sample using heavy metal stains. The samples for analysis in TEM is general prepared in special techniques like (i) Casting replica with a simple replica or a direct replica. Replica is made by applying a very thin film of 0.1% solution of Collodium to the sample surface, which is floated off and dried and then examined. Direct replica is made by vapour deposition with carbon or with heavy metal. The deposition is then floated off or obtained as a free film. However, sometimes, these are not suitable for structural analysis. The sequence of elements comprising a transmission electron microscope is analogous to that of a light microscope. It consists of (i) Electron gun with a Wehnelt cylinder, (ii) Electromagnetic lenses and (iii) Fluorescent screen. The image in the TEM is formed by monochromatic electrons with the aid of electromagnetic lenses in a high vacuum. The accelerating voltages is in between 60 and 120 kV, and sometimes as high as 200 kV. The wavelengths attainable are 0.005–0.002 nm, respectively. The point resolving power is about 0.2 nm. The image is mainly formed by the absorption of scattered electrons in the object. Owing to the strong interaction of electrons with matter, only small material thickness restricted to about 1.0 µm is allowed. TEM offer the best image resolution of all microscopy techniques.

Transmission electron microscopy techniques offer the best image resolution of all microscopy techniques and structural information relating to the following structures: (a) Morphological structure, (b) multicomponent structure (c) distribution of crystalline super lattices in semi-crystalline polymers, (d) globular structures, if any (e) surface defects after mechanical failure, (f) crystalline phase structures, (g) dispersed phase structures. However, this technique requires high capital expenditure, tedious, time consuming, skilled and complicated sample preparation methods. The specimen must be extremely thin, the order of 50 nm. The interpretation is also difficult and requires background knowledge. Transmission electron microscope provides high resolution and magnification (58–1,000,000×) and it enables extremely detailed examination of the internal structure of fibres, providing complementary information to light microscope and scanning electron microscope. TEM is ideally suited to research investigating the effect of new processes, as well as the routine diagnosis of fibre modification characteristics resulting from processing treatments such as bleaching, chlorination, flameproofing, mordanting or enzyme reaction. TEM is useful for studying fibre damage and any modifications caused by processing, whether caused by physical impact or chemical reaction with textile reagents.

The microscope consists of a source supplying a beam of electrons of uniform velocity, a condenser lens for concentrating the electrons on the specimen, a specimen stage for displacing the specimen which transmits the electron beam, an objective lens, a projector lens, and a fluorescent screen on which the final image is observed. For permanent record of the image, the fluorescent screen is replaced by a photographic plate or film. Modern TEM includes aberration correctors, to reduce the amount of distortion in the image, allowing information on features on the scale of 0.1 nm to be obtained. Monochromators may also be used which reduce the energy spread of the incident electron beam to less than 0.15 eV. Another type of TEM is the Scanning Transmission Electron Microscope (STEM), where the beam can be scanned across the sample to form the image. It is also possible to use a secondary electron detector and scan the electron beam over the sample to generate an image in the TEM very similar to that obtained in the SEM. This technique is called STEM.

A crystalline material interacts with the electron beam mostly by diffraction. The intensity of the transmitted beam is affected by the volume and density of the material through which it passes. The intensity of the diffraction depends on the orientation of the planes of atoms in a crystal relative to the electron beam – at certain angles the electron beam is diffracted strongly from the axis of the incoming beam, while at other angles the beam is largely transmitted. The sample can also be tilted to different angles to obtain specific diffraction conditions, and apertures placed below the specimen allow the user to select electrons diffracted in a particular direction. A high contrast image can be formed by blocking electrons deflected away from the optical axis of the microscope by placing the aperture to allow only unscattered electrons through. This produces a variation in the electron intensity that reveals information on the crystal structure, and can be viewed on a fluorescent screen, or recorded on photographic film or captured electronically. TEM is used to investigate the crystal structure. Crystal structure can also be investigated by High Resolution Transmission Electron Microscopy, also known as phase contrast imaging as the images are formed due to differences in phase of electron waves scattered through a thin specimen.

9.13.1 Limitations

There are a number of drawbacks to the TEM technique. Many materials require extensive sample preparation to produce a sample thin enough to be electron transparent. This makes TEM analysis relatively time consuming process. The structure of the sample may also be changed during the preparation process. Also the field of view is relatively small, so the region analysed may not be characteristic of the whole sample. There is potential that the sample may be damaged by the electron beam. The thin specimens used in TEM must be able to withstand the high vacuum present inside the instrument.

9.13.2 Microscopy sample preparation: microtomes and ultramicrotomes

For both transmission electron microscopy and transmitted light microscopy it is necessary to produce samples which are thin enough for either light (approximately 1/50th the diameter of a human hair) or electrons (approximately 1/500th the diameter of a human hair) to pass through. The first step is to embed the biological sample in a material which will support it while it is being cut. A resin is normally used to do this. The next step is to cut the sample into thin slices. This is done by using either a microtome or an ultramicrotome. These devices use special glass or diamond knives to produce the thin sections required. The TEM requires thin (around 100 nm) sections. Sample preparation can utilize either glass and/or diamond knives.

9.14   Scanning electron microscopy (SEM)

Scanning Electron Microscopy (SEM) is used exclusively for examination of sample surfaces. The electron beam scans across the specimen in a series of parallel lines. The scanning is continuous with line by line scanning. This produces images by detecting secondary electrons which are emitted from the surface due to excitation by the primary electron beam. The image can be recorded formed by collecting secondary electrons emitted from the surface by means of a suitable detector, i.e. scintillator and reproducing scintillator image. The resolution in SEM is less (~ 100 Å) than that obtainable in TEM. The main elements in SEM are: (a) Electron gun (heated cathode), (b) electromagnetic lenses for focusing electrons, (c) sample chamber with specimen stage and detectors and (d) electronic system to display image. The electrons emitted from the electron gun are focused on the object surface by means of a Wehnelt Cylinder and 2–3 electromagnetic lenses. The sweep generator control the electron beam scans. The secondary electrons (SE) and/or back scattered electrons (BE) emitted from the sample are monitored by suitable detectors. Signals emerging from the detectors with a video amplifier modulate the intensity on the screen. The magnification is usually 10 to 100,000 times, depending upon the scanned sample surface. The size of the scanned area determines the magnification of the image. The generated secondary electrons are collected and an image is formed. The image shows the surface structure and features of the sample in great detail. The detected signal is displayed as a TV type of image. Small samples of up to several millimeters can be investigated directly in the scanning electron microscope.

Generally, the TEM resolution is about an order of magnitude better than the SEM resolution, however, because the SEM image relies on surface processes rather than transmission it is able to image bulk samples and has a much greater depth of view, and so can produce images that are a good representation of the 3D structure of the sample. SEM technique is more advantageous than transmission electron microscopy. The technique can give three-dimensional images, great depth of field and focus, ease of specimen preparation and ease of interpretation. The image lacks internal structure and sometimes the specimen may be unstable to the applied vacuum and electron beam during investigation. Replica technique is used to avoid the instability of the material to high vacuum. It is necessary to coat the samples with chromium or carbon to provide a degree of electrical conductivity. In order to accomplish this, a chromium coater or a carbon coater should be used. Other samples can be polished (0.25 µm surface roughness) for analysis.

SEM is very much useful for study of polymer and fibre surfaces and as an aid to problem solving. At present SEM is mainly utilized to correlate the structure and properties of polymers and fibres. Scanning electron microscope facilities provide the requisite magnification suitable for the observation of the surface structure of fibres and textile products. Routine textile SEM applications include examination of fibre, yarn and fibre characteristics, fibre fracture morphology, and determination of the size and distribution of contaminants, powders or polymer depositions. Skills gained in textile applications have been recently used to study biopolymer structures.

9.15   Scanned probe microscopy (SPM)

In Scanned Probe Microscopy (SPM), a very fine tip is progressively scanned across the sample. This tip is attached to a cantilever and the deflection of the cantilever corresponds to the force between the sample and tip. During scanning the tip/sample force is kept constant by adjusting the height of the tip to accommodate changes in the height of the sample. In this way images of the surface topography are obtained. By measuring the lateral force required to move the tip across the sample the frictional properties of the material can be determined and by forcing the tip into the sample, the determination of nanomechanical properties such as the modulus and hardness of the sample is possible. Samples can be investigated in an air or liquid environment. In SPM no sample preparation is required. Occasionally samples may need to be embedded and sectioned.

Further readings

1. D. Campbell, R. A. Pethrick, J. R. White, *Polymer Characterization Physical Techniques*. Chapman and Hall, 1989.

2. *Encyclopedia of Polymer Science and Engineering*, Wiley, New York, 1986.

3. Linda C. Sawyer, David T. Grubb, Gregory F. Meyers, *Polymer Microscopy*, Springer, 2008.

4. R.S. Stein, *Newer Methods of Polymer Characterization*, R. Ke, (Ed.), Interscience Publishers, New York, 1964), p. 185.

Diffraction methods

10.1 Introduction

Diffraction refers to various phenomena which occur when a wave encounters an obstacle or a slit. These characteristic behaviours are exhibited when a wave encounters an obstacle or a slit that is comparable in size to its wavelength. Diffraction occurs with all waves like light waves, sound waves, x-rays, electrons and radio waves.

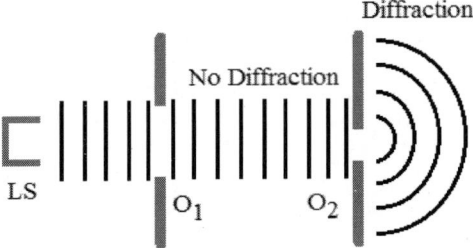

Figure 10.1 Diffraction.

Diffraction or scattering is a separate category of analytic techniques using electromagnetic radiation where the interference of radiation arising from structural features is observed. The basic principles of scattering and diffraction are the same, while the implementation of these principles is quite different. Diffraction is a coherent process and scattering is an incoherent process.

Diffraction methods provide detailed and concrete information regarding conformation of the single polymer chain, its orientation, its short and long range order in the solid and its relationship with neighbouring polymer chain. This includes qualitative and quantitative determination of the crystalline components in the semicrystalline polymers like fibres. There are two diffraction techniques, where different sources are used for diffraction. Those are : (a) x-ray sources of 1.0–2.0 Å and the phenomenon is X-Ray Diffraction (XRD), (b) electron sources of 0.01 Å and the phenomenon is Electron Diffraction (ED). These diffraction techniques provide overall information that must be interpreted by suitable molecular modelling. These techniques do not provide direct signals that arise from specific atomic groups in a specific structure.

10.2   X-ray diffraction

XRD is considered as an independent method for determining the structure and particularly the crystalline structure in polymeric materials like fibres. X-rays are electromagnetic waves of the order of 1.0–2.0 Å. When x-ray beam is incident on crystalline material, diffraction occurs and a number of diffracted rays appear in addition to the primary beam. The direction and intensity of the diffracted rays are recorded on a photographic plate. In the photographic plate, a dark spot appears at the point of incidence. The direction of diffracted rays can be calculated by means of Bragg's equation.

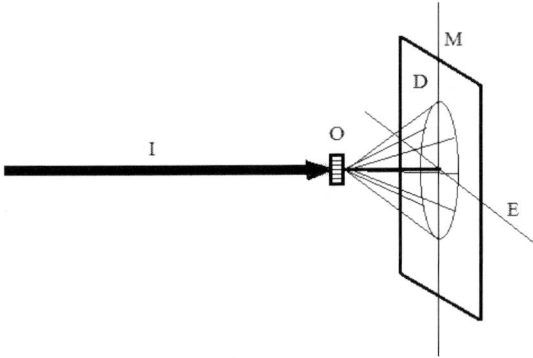

Figure 10.2   X-ray diffraction: I, Incident beam; O, Object; D, Diffracted pattern; M, Meridian; E, Equator.

Fibre diffraction is the only practical method of structure determination at the molecular level for filamentous systems. These systems range from simple polymers to complex assemblies including fibres and filaments. Some of the components do not crystallize, as their natural tendency is to form helical aggregates. Some components do crystallize but the crystals do not contain the intermolecular interactions that give the molecule its biological function. The defining difference between fibers and crystals is that, in fibers, the molecules are parallel to each other, but are randomly oriented about the fiber axis. In consequence, the diffraction pattern is cylindrically averaged. The combination of the cylindrical averaging and the inherent disorder common among natural filaments makes structural analysis difficult.

X-ray is used for determination of the crystal structure of crystalline polymers (size, shape of the unit cell, crystallite size, dimension and amount, orientation of the crystallites). X-ray structure is based on the analysis of the coherent diffraction of x-rays by regularly ordered three-dimensional periodic arrangements of atoms in the crystal. Crystalline materials give rise to sharp diffraction rings or peaks in accordance with the Bragg's law. Synthetic polymers and fibres never occur as single crystals. The diffraction pattern from polymers is almost always either a 'powder' pattern (polycrystalline) or a fiber pattern (oriented polycrystalline). The amorphous phase is a random distribution of atoms which scatter x-ray beam as a function of the particular electron density of each atom and the distribution of interatomic distances. Amorphous materials produce broad diffuse scattering of x-rays, as seen in Fig. 10.3(a). In general, the diffraction efficiencies to x-ray of the crystalline and amorphous regions are same. So XRD pattern of fibres or semicrystalline polymers usually have a wide amorphous halo along with sharp and well defined diffracted rings and arcs. The analysis gives valuable information on the configuration of macromolecules, crystallinity, crystallite size, lattice constants, crystallite orientation, molecular packing and order and amorphous structure. The postulation of x-ray analysis is that the polymer or the fibre exists as two definite phases, i.e. crystalline and amorphous. Each diffracts x-ray in its own characteristic pattern.

10.3   Information from XRD

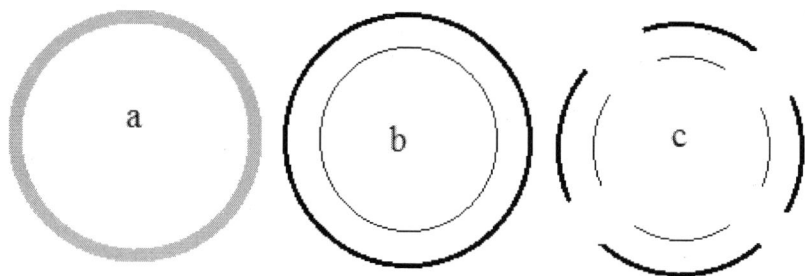

Figure 10.3   X-ray diffractogram of (a) amorphous sample, (b) crystalline sample, and (c) oriented sample.

The intensity distribution in a XRD due to the crystalline fraction is a superposition of the diffraction from many individual lattices of the crystallites. For a reflection of spacing 'd', the intensity is distributed on the surface of the sphere of radius '1/d'. This sphere is called the sphere of reflection. For crystallites with no orientation or random orientation, the distribution will be uniform, on the other hand, for the oriented crystallites, there will be fluctuations with higher and lower density [Fig. 10.3(a) and (b)].

For uniaxial orientation, the regions will be of cylindrical symmetry. Based on this, different parameters which can be analysed by XRD are:

(a) Parameters related to lateral packing order: The distribution of phases in the plane perpendicular to the fibre axis, the average extension of the regions with a particular type of order, the lateral spacing between chains in each phase, the average position of the chains inside the projected structure in a plane perpendicular to the fibre axis, the degree of orientation of the predominant phase.

(b) Parameters related to longitudinal packing order: The distribution of phases in the total sample, crystallizable impurities, amorphous impurities of high concentration, the overall crystallinity.

The detailed analysis of the x-ray diffractogram can indicate different aspects of the crystallites like its dimensions volume or weight fraction. To determine the above parameters, three scans are necessary.

(1) Equatorial transmissions scan obtained by placing the fibre axis perpendicular to the plane of the counter's movement (E in Fig. 10.2).

(2) Meridional transmissions scan obtained by placing the fibre axis in the plane of the counter's movement (M in Fig. 10.2).

(3) Disoriented transmissions scan obtained in the same 2θ range by rotating the sample with the aid of a rotary specimen holder, at approximately 120 rpm.

The difractogram scans are used to obtain for polymers and fibres at 2θ from 3° to 35°. Polymers are not highly absorbing to x-rays. XRD is a transmission experiment where the x-ray beam passes through the sample. This greatly simplifies analysis of diffraction spectra for polymers but requires somewhat specialized diffractometer. Typically the optimal thickness for a hydrocarbon polymer is 2 mm.

Four main features of XRD are of importance to polymer or fibre analysis. Those are:

1. Degree of crystallinity

2. Microstructure

3. Indexing of crystal structure

4. Orientation

10.4 Degree of crystallinity

Polymers are never 100% crystalline since the stereochemistry is never perfect, chains contain defects such as branches, and crystallization is highly rate dependent in polymers due to the high viscosity and low transport rates in polymer melts. A primary use of XRD in polymers is determination of the degree of crystallinity. The determination of the degree of crystallinity implies use of a two-phase model, i.e. the sample is composed of crystals and amorphous and no regions of semicrystalline organization. The integrated XRD intensity measures the volume fraction crystallinity, X_v. Other techniques like density gradient columns measure a mass fraction crystallinity X_w. The two fractions are related by the density ratios, where ρ_c is the crystalline density, ρ_s the bulk sample density and ρ_a is the amorphous density.

Postulations :

(1) A basic postulation of all x-ray methods is that the scattering efficiencies to x-rays of the crystalline and amorphous regions are the same. This statement should be further qualified by stipulating that the orientation of crystalline material present must be sufficiently randomized that all diffractions are equally probable. For this reason oriented films are unsuitable for x-ray crystallinity determinations, as some reflections are completely absent. This can be avoided by randomization of the sample.

(2) A second postulate is that the polymer exists as two definite phases, i.e. crystalline and amorphous phase. Each scattering has its own characteristic pattern. Purely amorphous, i.e. 100% amorphous pattern will exhibit a randomized pattern and 100% crystalline sample will exhibit a pattern containing all the crystallographic planes. One phase is embedded in another phase. Then, the scattering from any sample can be described as a weighted sum of the two regions, i.e. crystalline content and amorphous content.

(3) There are limits to the two-phase model, particularly for fairly disorganized polymer crystalline systems such as polyacrylonitrile.

(4) Most polymer systems are amenable to the two-phase model but the 2-phase model ignores interfacial zones where the density may differ from that of the amorphous.

10.4.1 Crystallinity percentage

The most simple and straightforward method is the estimation of crystallinity is in percentage with these assumptions. Here, the area under each region can be separated and the respective fraction can be estimated. Determination of X_c from the XRD pattern under the 2-phase assumption involves separation the diffraction pattern into three parts: (1) Crystalline, (2) Amorphous and (3) Background (incoherent scattering). The diffracted intensity is proportional to the amount of each of these contributions. The actual scattered intensity is related to a volume integral of the diffracted peak in the pattern. The amorphous scattering curve can be constructed by proportionally reducing the scattering curve of a completely amorphous sample until it matches the observed intensity at one or two particular diffraction angles (Fig. 10.4).

All the methods are based on the proportionality of the intensity diffracted by the amorphous and crystalline phases with the respective weight fractions, as per the following equation:

$$X_w (\%) = 100 \cdot A_c /(A_c + A_A)$$

where X_w is the weight fraction of crystalline area, A is the area under the respective phase, the subscript 'c' stands for crystalline fraction and 'a' for amorphous fraction. A number of methods are evaluated based on external comparison with justification and accuracy. Each procedure has its own advantages and disadvantages.

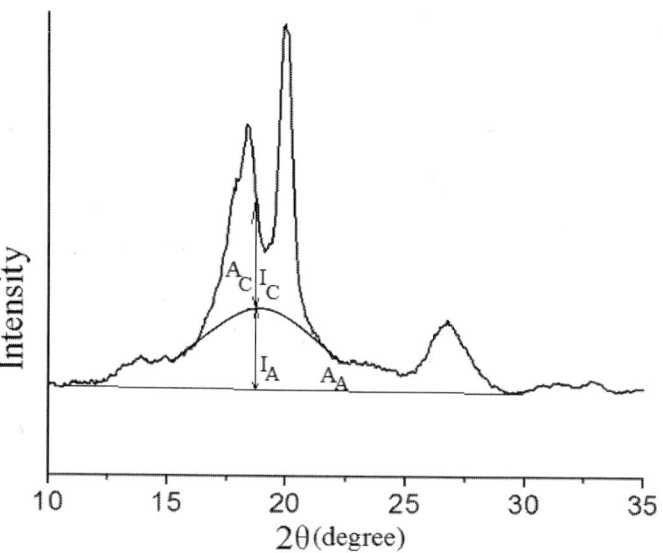

Figure 10.4   Measurement of crystallinity.

More accurately crystallinity can be calculated by means of the following equation:

$$X_c = \frac{\int\limits_{s_1}^{s_2} s^2 I_c(s)\,ds}{\int\limits_{s_1}^{s_2} s^2 I(s)\,ds}$$

where $s = 2/\lambda \cdot (\sin \theta)$, λ, the wavelength, θ, Bragg's angle. The equation can be further simplified as

$$X_c = \frac{\sum\limits_{s_1}^{s_2} s^2 I_c(s) ds}{\sum\limits_{s_1}^{s_2} s^2 I(s) ds}$$

The volume fraction degree of crystallinity is given by the ratio of the integral of the crystalline diffraction intensity over the total coherent scattering, i.e. after subtracting the incoherent scattering. This equation requires a mathematical computation of the intensity obtained by the sample and its crystalline component.

10.4.2 Crystallinity index

The estimation of absolute crystallinity is different. XRD patterns are recorded on the most 'amorphous' and the most 'crystalline' samples obtainable and these are taken as standards of zero and '100' crystalline index.

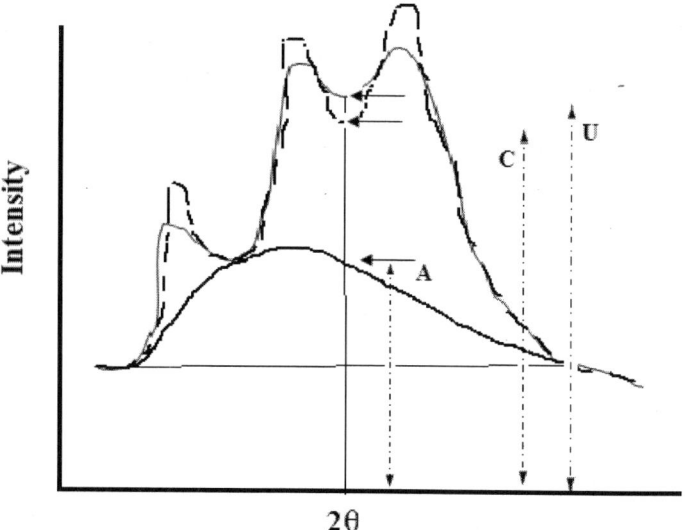

Figure 10.5 Measurement of crystallinity index.

Instead of separation of the amorphous content from the scattering with estimation of the fraction of area, crystallinity index can be calculated by means of a standard (100%) crystalline sample and 100% amorphous sample. The unknown sample and both the standard samples will be plotted. The assumption is that the intensity on either side of the amorphous halo is independent of crystallinity.

The corrected intensity for the amorphous standard at a given scattering angle was subtracted from that of the crystalline standard at the same scattering angle. Point by point differences should be obtained at each scattering angle from 5° to 50°. The amorphous standard differences were then corrected with the crystalline standard and amorphous differences. This method is more suitable for cotton as well as polyester samples. The main advantage of this method is that it avoids drawing or plotting of the amorphous curve and separation of crystalline area and amorphous area like other methods. However, it does not show absolute crystallinity. The crystalline sample and unknown sample have to show the same peaks in same positions. For this reason, this method is not suitable for nylon samples or polyethylene. Crystallinity index can be calculated by any of the two methods, viz. (a) integral method and (b) correlation index method.

10.4.2.1 Integral method

The differences in the corrected intensity readings for the cotton amorphous standard and the crystalline amorphous standard were summed without regard for sign and the ratio of the former to the latter sum taken as a measure of crystalline content. In effect, this method compares the included area between the curves for the crystalline and amorphous standards.

The differences $(U - A)$ and $(C - A)$ are accordingly summed without regard for sign to provide an estimate of crystallinity. The crystalline index (X_i) may be represented as follows:

$$X_i(\%) = \frac{\sum_{2\theta}(U - A)_{2\theta}}{\sum_{2\theta}(C - A)_{2\theta}} \times 100$$

In the integral method, the incoherent scattering is also removed in the differences $(U - A)_2\theta$ and $(C - A)_2\theta$. The multiplicative correctness is also removed. This is easier of the two methods and can be simply adopted for routine works.

10.4.2.2 Correlation index

This method calculates the correlation index between of the unknown sample by means of the standard samples. The slope of the regression line provides an estimate of crystallinity and the correlation co-efficient an estimate of the error associated with the slope.

Designating the average crystalline and amorphous data as C and A, respectively, the differences $(C - A)_2\theta$ was taken at every value of scattering angle 2θ from 5° to 50°. The average amorphous data were then subtracted at corresponding values of 2θ from the data for each cotton sample (U) to form $(U - A)_2\theta$. The correlation of the values of $(C - A)_2\theta$ and $(U - A)_2\theta$ at corresponding scattering angle then gave a regression line whose slope was taken as an index of the crystalline content, or the departure of cotton from the amorphous standard and a correlation co-efficient as a measure of the spread of the data about the regression line.

Letting $Y_2\theta = (U - A)_2\theta$ and $X_2\theta = (C - A)_2\theta$ for these parameters at each value of the scattering angle 2θ, the crystalline index is defined by the slope of the least squares regression of $Y_2\theta$ on $X_2\theta$ as follows :

$$X_i(\%) = \frac{\sum_{2\theta} X_{2\theta} \cdot Y_{2\theta} - (\sum_{2\theta} X_{2\theta} \cdot \sum_{2\theta} Y_{2\theta} / N)}{\sum_{2\theta} X_{2\theta}^2 - (\sum_{2\theta} X_{2\theta}^2 / N)} \times 100$$

where N is the total number of pairs of observations. The spread of the data about the regression line is estimated by $1 - \gamma^2$, where γ is the correlation co-efficient

$$\gamma = \frac{\sum_{2\theta} X_{2\theta} \cdot Y_{2\theta} - (\sum_{2\theta} X_{2\theta} \cdot \sum_{2\theta} Y_{2\theta} / N)}{[\sum_{2\theta} X_{2\theta}^2 - (\sum_{2\theta} X_{2\theta}^2 / N) \sum_{2\theta} Y_{2\theta}^2 - (\sum_{2\theta} Y_{2\theta}^2 / N)]}$$

$$\gamma = \frac{N \sum_{2\theta}(C - A)(U - A) - \sum_{2\theta}(C - A) \cdot \sum_{2\theta}(U - A)}{N \sum_{2\theta}(C - A)^2 - [\sum_{2\theta}(C - A)]^2}$$

where C is the intensity diffracted for the crystalline standard at diffraction angle 2θ, A is the intensity scattered for the amorphous standard at diffraction angle 2θ, U is the intensity diffracted and scattered for the sample being evaluated at diffraction angle 2θ.

The correlation method has several advantages. The multiplicative correction factors for polarization and spectrometer geometry are functions of the scattering angle. These are eliminated by taking the ratio of the individual 2θ values of $Y_2\theta/X_2\theta = (U - A)_2\theta/(C - A)_2\theta$ in the correction. Secondly, if a unit mass of cellulose produces C, A and U scattering curves, the incoherent scattering will be the same for all these three and differences $(U - A)_2\theta$, $(C - A)_2\theta$ are the coherent parts of the scattered intensity each.

10.5   Microstructure

Crystallite size in polymers is usually on the nano-scale in the thickness direction. The size of crystallites can be determined using variants of the Scherrer equation. The Scherrer equation relates the size of crystallites of sub-micrometer particles in a solid to the broadening of a peak in a diffraction pattern. The Scherrer equation can be as

$$\tau = \frac{K\lambda}{\beta \cos \theta}$$

where τ is the mean size of the ordered (crystalline) domains, which may be smaller or equal to the grain size; K is a dimensionless shape factor, with a value close to unity. The shape factor has a typical value of about 0.9, but varies with the actual shape of the crystallite; λ is the x-ray wavelength; β is the line broadening at half the maximum intensity (full width at half maximum, FWHM), after subtracting the instrumental line broadening, in radians. This quantity is also sometimes denoted as $\Delta(2\theta)$; θ is the Bragg angle. Calculation of FWHM is shown in Fig. 10.6. This formula is used in the determination of size of particles of crystals in the form of powder.

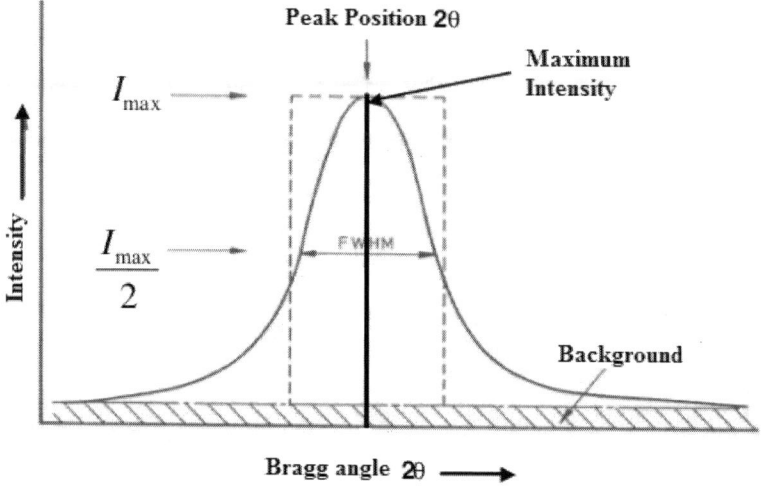

Figure 10.6   Calculation of full width at half maximum (FWHM).

10.6   Indexing of crystal structures

Indexing of crystal structures can be done to determine the cell parameters and the crystal lattices. It is important that the powder pattern should be used with a single phase, displaying little or no orientation and should contain narrow peaks that have little overlaps with neighbouring peaks. The following procedure should be adopted for indexing.

1.   The angle measured corresponds to 2θ, where θ is the angle in the Bragg law.

2.   Using the Bragg's law and the wave length of the incoming radiation, calculate d_{hkl} for all available peaks.

3.   Pick one as a reference $d(h_1k_1l_1)$ and take the ratio of $d^2(h_1k_1l_1)/d^2(h_2k_2l_2)$.

4.   Multiply this ratio by an integer of increasing magnitude until all are whole numbers. Or if the crystal structure is known, multiply until all

planes are consistent with that structure. This is equivalent to applying extinction rules for the crystal structure type

5. Using d and a single set of $\{hkl\}$ planes, calculate

$$a = d \cdot x \langle (h^2 + k^2 + l^2) \rangle^{1/2}$$

Please note that the same value of a should be obtained regardless of which set of $\{hkl\}$ is used.

10.7   Electron diffraction (ED)

ED is associated with interference processes that occur when electrons are scattered by atoms to form diffraction patterns. The wave character of electrons is created by the phenomena of interference. For this reason, the diffraction of electrons follows the rules of quantum mechanics. Because of the dependence of the diffraction pattern on the distances between the atoms, ED is an important tool for the study of the structure of crystals and of free molecules, analogous to the use of x-rays.

10.8   Electron interaction with matter

Electrons are charged particles and interact with matter through the Coulomb forces. This means that the incident electrons feel the influence of both the positively charged atomic nuclei and the surrounding electrons. The wavelength of an electron is given by the de Broglie equation

$$\lambda = h/\rho$$

Here h is Planck's constant and ρ the momentum of the electron. The electrons are accelerated to the desired velocity 'υ':

$$\upsilon = (2e \cdot v/m)^{1/2}$$

where m is the mass of the electron; e is the charge on electron; v is the accelerating voltage in volts. The electron wavelength is then given by

$$\lambda = h/\rho = h/m\upsilon = h/(2m \; e \cdot v)^{1/2}$$

The wavelength of the electron will follow the above principle. This equation can be simplified, as e and m are practically constants. The equation will be

$$\lambda = (3/2v)^{1/2}$$

For $v = 40$ kV, the value of λ will be 0.006 nm.

Applying Bragg's law like XRD

$$\lambda = 2 \cdot d \cdot \sin \theta$$

when the value of λ is compared for XRD and ED, for same value of d, the value of θ must be smaller.

If a beam of electron is trained on crystalline substance, it will give an ED pattern on a photographic plate, similar to XRD pattern. So the geometry of ED is same as that of XRD. However, the wavelength of ED is smaller than that of XRD.

The intensity of ED is higher than that of XRD owing to strong interaction. So the technique requires

1. Extremely fine thickness of the sample (10–100 nm).

2. Little amount of sample (10^{-12} to 10^{-13} g).

3. Short exposure time.

4. Precaution for small lattice distortion.

The diffraction measurements are usually carried out in a transmission electron microscope (TEM). The methods of treating experimental data for XRD and ED are almost identical. ED is used to study the structure of crystalline and amorphous substances like XRD analysis.

10.9 Techniques of electron diffraction

According to energy $E = eV$ (where e is electron charge and V is potential difference), two major techniques of structure analysis with electron beams are distinguished: (a) low-energy electron diffraction (LEED) [$E = 5$–500 eV] and (b) high-energy electron diffraction (HEED) [$E = 5$–500 keV].

10.9.1 Low-energy electron diffraction

LEED is used mainly for the study of the structure of single-crystal surfaces and of processes on such surfaces that are associated with changes in the lateral periodicity of the surface. A monochromatic, nearly parallel electron beam, of 10^{-4} to 10^{-3} m (4×10^{-3} to 4×10^{-2} in.) in diameter, strikes the surface, usually at normal incidence. The elastically backscattered electrons are separated from all other electrons by a retarding field and detected with a suitable movable collector or, more frequently on a hemispherical fluorescent screen with the crystal in its center. The intensity of the diffraction spots can be measured as a function of the energy of the incident electrons.

The most important contribution of LEED is to the understanding of the structure of many adsorption systems, mainly of gases on metals, or metals on other metals and semiconductors. LEED is also helpful for the kinetics of the adsorption and desorption processes as well as changes in the adsorption layer upon heating.

10.9.2 High-energy electron diffraction

HEED is used mainly for the study of the structure of thin foils, films, and small particles (thickness or diameter of 10^{-9} to 10^{-6} m), of molecules, and also of the surfaces of crystalline materials. A monochromatic, usually nearly parallel, electron beam with a diameter of 10^{-3} to 10^{-8} m is incident on the target. The forward-scattered electrons are detected by means of a fluorescent screen, a photoplate, or some other current-sensitive detector, usually without the inelastically scattered electrons being eliminated.

In scanning HEED the diffracted electrons are not recorded on photographic film but are directly measured electronically with sensitive detectors. By moving the detector across the diffraction pattern or by deflecting the diffracted electrons across a stationary detector (scanning), the intensity distribution in the diffraction pattern can be displayed quantitatively on an XY recorder.

ED is most frequently used to study the crystal structure of solids. These experiments are usually performed in a TEM or a scanning electron microscope as electron backscatter diffraction. In these instruments, the electrons are accelerated by an electrostatic potential in order to gain the desired energy and wavelength before they interact with the sample to be studied.

10.10 Transmission electron microscope

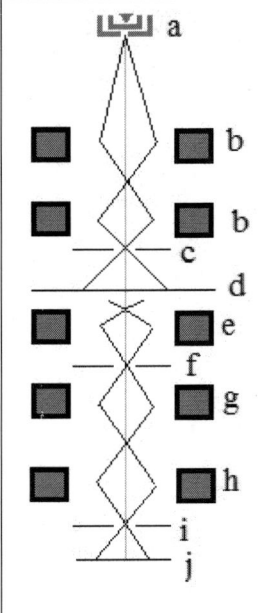

Figure 10.7 Transmission electron microscope: (a) Electron gun, (b) Condenser lens, (c) First aperture, (d) Specimen, (e) Objective lens, (f) Second aperture, (g) Intermediate lens, (h) Projection lens, (i) Final aperture, (j) Final lens.

The schematic diagram of transmission electron microscope is shown in Fig. 10.7. In this microscope, an electron beam from an electron gun is transmitted through an ultra-thin section of the microscopic object and the image is magnified by the electromagnetic fields. It is used to observe finer details of internal structures of microscopic objects. The specimen to be examined is prepared as an extremely thin dry film or as an ultra-thin section on a small screen and is introduced into the microscope at a point between the magnetic condenser and the magnetic objective.

10.11   Scanning electron microscope

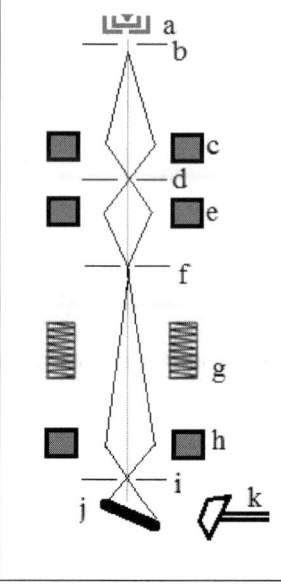

Figure 10.8 Scanning electron microscope: (a) Electron gun, (b) First aperture, (c) First condenser lens, (d) Second aperture, (e) Second condenser lens, (f) Third aperture, (g) Scanning coil, (h) Final condenser lens, (i) Final aperture, (j) Specimen, (k) Collector

The schematic diagram of transmission electron microscope is shown in Fig. 10.8. In a scanning electron microscope, the specimen is exposed to a narrow electron beam from an electron gun, which rapidly moves over or scans the surface of the specimen. This causes the release of a shower of secondary electrons and other types of radiations from the specimen surface.

The intensity of these secondary electrons depends upon the shape and the chemical composition of the irradiated object. These electrons are collected by a detector, which generates electronic signals. These signals are scanned in the manner of a television system to produce an image on a cathode ray tube (CRT). The image is recorded by capturing it from the CRT. Modern variants have facility to record the photograph by digital camera. This microscope is used to observe the surface structure of microscopic objects.

10.12   Scanning and transmission electron microscope

It has both the facility of transmission and scanning and so referred as scanning and transmission electron microscope functions.

Further readings

1.   Claudio De Rosa and Finizia Auriemma, *Crystals and Crystallinity in Polymers: Diffraction Analysis of Ordered and Disordered Crystals*, Wiley, 2013.

2.   *Encyclopedia of Polymer Science and Engineering*, Wiley, New York, 1986.

3. I. M. Ward (Ed.), *Structure and Properties of Oriented Polymers*, Springer, Netherlands, 1997.

4. H. Tadokoro, *Structure of Crystalline Polymers*, Wiley, New York, 1979.

5. Leroy E. Alexander, *X-Ray Diffraction Methods in Polymer Science*, Wiley-Interscience, New York, 1970.

6. Oliver H. Seeck and Bridget Murphy, *X-Ray Diffraction: Modern Experimental Techniques*, CRC Press, 2015.

7. Emil Zolotoyabko, *Basic Concepts of X-Ray Diffraction*, Wiley, 2014.

8. E. J. Mittemeijer and U. Welzel (Eds.), *Modern Diffraction Methods*, Wiley, 2012.

9. H. P. Klug and L. E. Alexander, *X-Ray Diffraction Procedures for Polycrystalline and Amorphous Materials*, , Wiley Interscience, 1974.

10. L. R. Alexander, *X-Ray Diffraction Methods for Polymer Science*, Kreiger Publication, 1979.

11. R. S. Stein, *Newer Methods of Polymer Characterization*, R. Ke (Ed.), Interscience Publishers, New York, 1964, p. 185.

Scattering techniques

11.1 Introduction

When a beam of radiation impinges on a sample, a portion of the radiation is scattered. This scattered fraction depends on both the nature of the radiation and the composition of the sample. The scattered ray becomes out of phase and they interfere. The amount of interference depends on scattering angle. The intensity of the scattering drops with increasing the scattering angle at a rate that depends upon particle size. For bigger particles, the rate of fall of intensity is greater. Different radiations like light, neutron and x-rays are used as sources for scattering studies. The wavelength ranges and thus the smallest structure that can be studied are:

Light	:	2000 Å–10,000 Å
Neutrons	:	0.1 Å–15 Å
X-ray	:	0.5 Å–5 Å

All the scattering experiments have certain features in common. A radiation source provides a parallel monochromatic beam of radiation. It passes through the sample. A detector measures the scattered intensity as a function of scattering angle and Bragg's angle. Small angle-scattering techniques are employed to measure, with sub-nm precision, pattern shape, dimensions and orientation for structures created in periodic arrays. Due to the nano-scale size of polymer crystallites, small-angle scattering is intense in semicrystalline polymers and a separate field of analysis based on diffraction at angles below 6° is used to study the structure of polymers and fibres.

11.2 Small-angle x-ray scattering (SAXS)

X-ray interference effects result from the variations in electron density from one point to another in the material. Particularly inhomogeneities of colloidal dimensions generate x-ray scattering and interference effects at very small angles, typically when 2 is less than 2° with the wavelength of 1.54 Å.

This SAXS has no dependence on inhomogeneities of atomic dimensions that give rise to wide-angle x-ray diffraction. The scattering of x-rays is related to the fluctuations of the electron density. X-ray sources are usually conventional sealed anode or rotating anode tubes. The x-ray source can be monochromatized by filtration. Collimation can be achieved either by slit or by pinhole. SAXS experiments are used to analyse the macrostructure of materials on a scale of about 1–200 nm. The studies may be divided into those of disoriented systems, analysed by statistical procedures and those from structured systems in which scattering is interpreted in terms of a model structure. SAXS provides morphological information like nature, size or shape of voids or crystalline regions, thickness and spacing of crystallites.

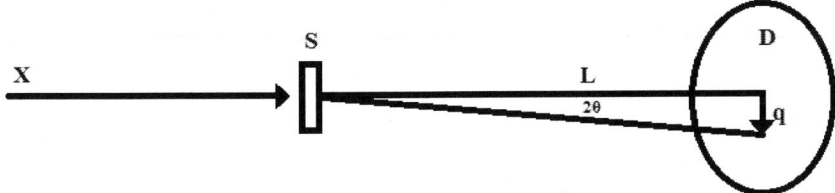

Figure 11.1   Small angle x-ray scattering: X, monochromatic x-ray beam; D, Detector; S, Sample; L, Distance, usually large between sample and detector; $q = 4\varpi \cdot \sin\theta/\lambda$, where θ is the Bragg's angle and λ is the wavelength of x-ray.

X-rays are scattered by electrons in matter and diffraction occurs between x-rays scattered by electron clouds that surrounded the various atomic centres. X-ray interference effect results from the variations in electron density from one point to another in the material. If the atoms are regularly arranged according to the space lattice, the scattering angle 2θ is related to the interplanar spacing by the Bragg equation, $n\lambda = 2d \sin\theta$. This equation indicates the existence of a reciprocal relationship between interatomic distances and $\sin\theta$. Due to reciprocal nature of the relationship, only scattering from gross or large-scale structures is present in small-angle region. This is referred as SAXS. It is a branch of x-ray analysis and emerged as a powerful tool for the study of macromolecular structure and morphology. The scattered intensity of a crystal composed of two components and it may be written as

$$I = I_1(\text{diffuse}) + I_2(\text{Bragg})$$

The diffuse scattering I_1 contains information on the near range order between units of the same kind whereas I_2 represents Bragg's scattering of the mean crystal lattice.

11.2.1 Instrumentation

By its very definition, SAXS requires acquisition of scattered intensity which is very close to the primary beam. For this, there are three basic requirements for the instrumentation. Those are:

(1) A highly collimated beam with small divergence to avoid distortion of intensity data.

(2) Elimination of parasitic scattering for obtaining true relative intensity data.

(3) A highly special purity of the primary beam.

So an intense source of x-ray is desired for SAXS work. High powered sealed tubes with or without rotating anode targets with 2–60 kW, 40–60 kV and 50–1000 mA facilities are generally used as x-ray source. Further the primary beam is also collimated for dictation of scattering at the smallest angle and to avoid the smearing effect. The collimation can be done either by pinhole collimators; slit collimators or by annular collimators. When using CuKα radiation with wavelength 1.542 Å, the scattering angle 2θ is in the range 2°

11.2.2 Utilization

In general, because of the reciprocity between interatomic distance and $\sin\theta$, inhomogeneities of colloidal dimensions generate x-ray scattering and interference effects at very small angles, typically less than 2° with wavelength of CuKα of 1.542 Å. This small-angle scattering has no dependence on the inhomogeneities of atomic dimensions that give rise to the wide angle diffraction. The concentrations of electrons on the atomic sites constitute the basis for the description of crystal structure. But in case of SAXS continuous distribution of electrons within the unit cell is the basis. The fluctuations of electron density over distances of 30–1000 Å determine the nature of small angle scattering. Therefore, perfect single crystals pure phases in general and other homogeneous substances do not scatter x-rays at very low angles. The two kinds of inhomogeneity that is most likely to be responsible for SAXS from solid polymers. Those are: (i) Alternation of crystalline and amorphous regions, which in general will possess different atomic densities and (ii) the presence of microvoids dispersed throughout the polymer matrix.

SAXS is used to investigate density fluctuations with characteristic dimensions on the order of tens to thousands of angstrom units. The majority of SAXS studies of polymers are on the two phase microstructure of semicrystalline polymers. Information which can be obtained from SAXS will depend upon particular specimen or scattering system under investigation.

If accurate data is obtained and the more direct quantitative theoretical techniques are applied, then the following types of information can be obtained concerning the macromolecular and morphological structure of the system. The information obtained from SAXS are: (a) direct 'particle' size, (b) macromolecular structure of the particle, (c) particle–particle correlation function, (d) matrix–particle correlation function, (e) lattice distortion parameter, (f) the macrolattice, (g) distribution of particle size, (h) orientation parameters between particles, (i) void structure, void–void correlation and void–particle correlation, (j) segregate macromolecular effects from molecular effects in the high angle, (k) heterogeneity and anisotropy, (l) electron density fluctuation.

SAXS is helpful to investigate the structure, both for amorphous and crystalline polymers. The information available from this scattering can be summarized as per the information given below:

(a) Amorphous polymers

(i) Heterogeneities in bulk amorphous polymers: Size, shape and concentration of heterogeneous particles.

(ii) Homogeneous amorphous polymers: Thermal density fluctuations.

(iii) Two phase systems: Size, shape and mutual arrangement of the two or more phases.

(iv) Structure caused by deformation: Morphological fine structure of deformation induced faults.

(b) Semicrystalline polymers

(i) Solution grown crystals : Thickness and thickness distribution of the crystal platelets, structure of the crystal surface, chain folding, crystallization kinetics and crystal growth.

(ii) Isotropic melt-crystallized polymers: Crystallinity, long spacing, specific inner surface, morphologcial inner structure in the range of 30–3000 Å. Size, shape and arrangements of crystalline and amorphous regions.

(iii) Oriented semicrystalline polymers: Crystallinity, long spacing, specific inner surface, morphologcial inner structure in the range of 30–3000 Å. Size, shape and arrangements of crystalline and amorphous regions, effect of orientation and mechanism of plastic deformation.

(iv) Crystallization, annealing and melting: Crystallization kinetics, changes of morphology by annealing and changes of morphology in the melting range.

(v) Effect of crystallization condition on the phase-transition temperature.

(vi) Change of morphology by stress and strain.

11.2.3 Structural information from SAXS

Polymeric materials show two types of scattering at small angles – diffuse scattering, which gives a halo and discrete diffraction resulting in a ring like scattering or streaks. Diffuse or void scattering corresponds to the formation of voids, like that of in solvent induced crystallization, while the polymer is in a swollen state. Diffuse scattering in solid polymers can be interpreted in terms of microvoids or intercrystalline space, and quantified in terms of the void size and the relative amount of the void volume present. Discrete SAXS provides information about the morphology of the system, the 'long spacing' providing data about the crystalline and the intercrystalline thickness, their sum and about their uniformity and shape. A ring scattering corresponds to a spherically symmetrical collection of crystallites as in spherulite. A discrete scattering corresponds to stacked lamellar crystals.

11.3   Small-angle light scattering (SALS)

SALS is an important technique in the study of morphology and crystallization behaviour of semicrystalline polymer. This scattering utilizes radiation of wavelengths in the range of visible light (4000–8000 Å). Structural aggregates of the order of 0.1 µm can be investigated and important knowledge regarding the structure of the material may often be acquired from the distribution of the scattered light. The scattering of visible light is related to the fluctuation of the refractive index or polarizability of the sample and its anisotropy. This polarizability is related to the mobility of the electrons, which is affected by the molecular structure of the sample. Lasers are generally used for the source of light as it provides extremely intense parallel and monochromatic sources. The scattered light after it passes through the sample can be recorded on a photographic film. A film of approximately 100 µm thick is generally used as the sample, which will provide sufficient scattering and it will transmit 90–95% of incident light.

Much of the work on light scattering from solid polymers has focused on studies on crystalline spherulite polymers. Spherulites are structures of semicrystalline polymers, which are in the size range of 1–100 µm. There are two types of pattern obtained from SALS. The pattern obtained, when the polarizer and the analyser is crossed, known as Hv pattern. All the Hv patterns are generally of the four leaf clover type. They exhibit a maximum in the light scattering intensity in each of the lobes, which is characteristic of spherulitic scattering.

On the other hand, the pattern obtained by parallel placement of both the polarizer and analyser is known as Vv pattern. Vv pattern (parallel polars) scattering arises from the density fluctuations as well as anisotropy, whereas the Hv arises only from anisotropy. Both are comparable in the absence of density fluctuations.

11.3.1 Theory of small angle light scattering

The theory of light scattering and the analysis is based on two types of approaches or thesis. Those are (a) Model approach and (b) Statistical approach. The model approach involves calculating the scattering amplitude arising from all the volume elements constituting some structural units such as the sphere or a disc, whereas in the statistical approach, a structure is described in terms of correlation function in density and orientation. In the model approach, the scattering region of the polymer is represented by an idealized structure and the scattering pattern is calculated from this structure. The polymer is idealized as an isolated anisotropic sphere, a disk or sometimes anisotropic sheaf or rods. The theoretical predictions differ from the experimental results because of complications arising from several factors such as inter-spherulitic interference, truncation among spherulites, internal disorder within the spherulites and multiple scatterings. The calculations are then modified to account for the deviation from ideality to the actual systems.

The scattering element has radial and tangential polarizabilities. The direction of the scattered ray is described in terms of scattering angle θ and azimuthal angle α. The scattered intensity can be calculated by averaging the square of the scattering amplitude. This amplitude is the sum of all amplitude contributions from all of the volume elements of the scattering objects. The intensity can be of two types, i.e. Hv pattern and Vv pattern. Hv pattern can be obtained by placing the polarizer and analyser perpendicular. On the other hand Vv pattern can be obtained by the parallel arrangement of polarizer and analyser. Complete light scattering patterns can now be easily obtained by calculating Hv and Vv intensity for several values of scattering angle θ and azimuthal angle α. The intensity depends upon the anisotropy of the spherulite and varies with $\sin^2\alpha \cos^2\alpha$ which gives a four leaf clover pattern with scattering maximum occurring at odd multiples of $\alpha = 45°$. The scattering angle at these maxima is a function of the spherulite radius given as:

$$U\mathrm{max} = 4\cdot\pi(R/\lambda)\cdot\sin qm/2 = 4.1$$

where λ is the wavelength of light in air, usually 6328 Å.

From this equation, the value of R is given as:

$$R = 1.025\lambda/p \cdot \sin qm/2$$

This equation is generally used to determine the shape and size of the spherulite in terms of radius R.

The Vv pattern depends upon the polarizability of the surroundings as well as on the anisotropy of the spherulites. The amplitude consists of two components. The first component depends on α, polarizability of the surrounding medium but is angularly independent. The second component depends on anisotropy and $\cos^2 \alpha$ giving rise to two-fold symmetry. The Hv scattering at small angles depends only upon fluctuations in the magnitude and optic axis orientation of anisotropic regions, where as the Vv scattering also depends upon fluctuations in the density or average polarizability. The Vv intensity maximum arises from the latter contribution. The origin of this average polarizability fluctuation is the difference between the average polarizability of the spherulite and of its surrounding material.

11.3.2 Instrumentation

The instrumentation of SALS is very simple. Also, exposure to light is not as detrimental to as that of x-ray. For polarized light, laser beam will be required. Once a laser beam is available, it is very easy to set up the instrument for SALS. Apart from laser beam, the sample holder and the receiver will form the set-up. The whole set-up should be present in a dark room to avoid any interference of light. A schematic diagram of small angle light scattering instrument is shown in Fig. 11.2.

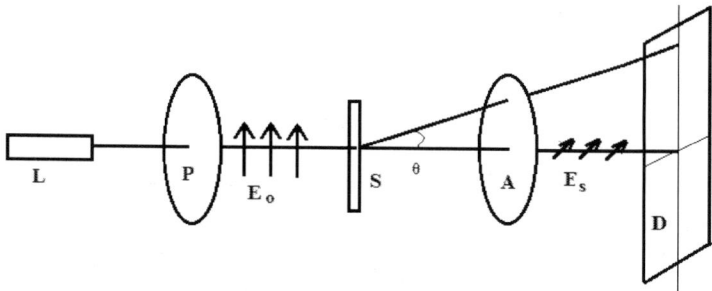

Figure 11.2   Small angle light scattering: L, Laser; P, Polarizer; S, Sample; A, Analyser, D, Detector.

It consists of the following parts: (i) Light source, (ii) Polarizer, (iii) Shutter, (iv) Sample holder, (v) Analyser, (vi) Camera/detector and (vii) Base. The base can be graduated to measure the distance between the sample and the camera. The slider of the bench is used to hold the different parts. For the light source, either a He-Ne gas laser of 5 mW output or a mercury arc light source with a pinhole and lens optics to render light parallel and filters for monochromatization can be used. For a light source, longer exposure times up to several hours may be required. So lasers provide coherent plane polarized monochromatic light without any optical lights and thus these are ideally light sources. The circular aperture is used to remove non-coherent uncollimated light. The polarizer polarizes the light in a parallel direction. The shutter is a mechanical switch to cut off the light, whenever needed. The sample holder holds the sample in a perfect manner such that the surface of the sample is perpendicular to the film. The analyser performs the same function as that of the polarizer. The scattered light can be collected on the flat plate camera.

P and A are two polarizers, where A is generally termed as analyser. The pattern obtained is designated as an Hv pattern when the polarizer P is vertical and the analyser A is horizontal and Vv when both the polarizer and analyser are vertical. The scattering depends upon the scattering angle θ between the incident and scattered ray and the azimuthal angle $\propto$. The range of θ which may be recorded depends upon the sample to film distance 'd' since tan $\theta = (X/d)$, where X is the distance from the centre of the photographic film to the point where it is recorded.

11.4   Small-angle neutron scattering

Neutrons are unchanged particles which may interact with a specimen by nuclear interactions. So neutron scattering is a nuclear property that varies with composition of the nucleus and is independent of the chemical nature of the species within which these nuclei reside. Neutron sources are steady state nuclear reactors giving a black body distribution of neutron velocity. Monochromatization is accomplished either by a velocity selector or else by diffraction from a crystal. Collimation can be achieved through pinholes and beam guides. The experimental set-up is similar to x-ray diffraction. The neutron scattering is applied to the study of conformation of chains in crystalline polymer or in dilute solutions, molecular dimensions of amorphous polymer, chain folding in polymer crystals, orientation and polymer blends. This technique helps to confirm random coil model in molten and in solid amorphous polymer. Neutrons are scattered by the atomic nuclei through the strong nuclear forces. In addition, the magnetic moment of neutrons is non-zero, and they are therefore also scattered by magnetic fields.

Further readings

1.   D. Campbell, R. A. Pethrick, J. R. White, *Polymer Characterization Physical Techniques*, Chapman and Hall, 1989.

2. L. R. Alexander, *X-ray Diffraction Methods for Polymer Science*, Kreiger Publication, 1979.

3. R.S. Stein, *Newer Methods of Polymer Characterization*, R. Ke (Ed.), Interscience Publishers, New York, 1964, p 185.

4. A. Giunier, G. Fournet, C. Walker, and K. Yudowich, *Small-Angle Scattering of X-ray*, Wiley, New York, 1955.

12.1 Introduction

Spectroscopy is the analysis of the interaction between matter and any portion of the electromagnetic spectrum. Spectroscopic methods play an important role in the structural analysis of polymers and fibres. Spectroscopy is defined as the science of interactions between electromagnetic radiation and matter such that energy is absorbed or emitted according to Bohr frequency conditions:

$$\Delta E = h \cdot v \qquad (12.1)$$

where ΔE is the energy difference between two successive energy levels; h, Planck's constant and v, the frequency. All forms of spectroscopy give spectra that may be described in terms of frequency, intensity and shape of spectral lines or bands. These observable properties depend on molecular parameters of the system.

Any form of spectroscopy give spectra that may be described in terms of frequency, intensity and shape of spectral line of bands. These observation properties depend upon the molecular parameters of the system. The study of molecular structure by spectroscopy depends primarily upon the existence of vibratory motions of atoms within the molecule. The molecule can be said to resemble a system of balls of varying masses, corresponding to the atoms of a molecule, and springs of varying lengths, corresponding to the chemical bonds of a molecule. These motions depend on the nature and arrangement of constituent atoms. Radiant energy falling upon matter is affected by the presence of such motions. A study of the behaviour of radiation falling upon matter is thus capable of giving indirect, but very valuable information on molecular structure. So the structure can provide the following information:

a.   Chemical structure

b.   Tacticity or configuration of the polymer

c.   Chain conformation

d.   Crystallinity

The spectroscopy yields precise information at the molecular level. Influences due to surrounding groups are evidenced by disturbances. Depending upon the disturbances and the interactions causing them, the structural information can be derived. The method involves investigation of the monomers, oligomers, single crystals, amorphous spectroscopy of the polymers. In order to get maximum useful information, other methods and polymer structure and properties should be combined with polymer spectroscopy. Different spectroscopy are indicated in the below table.

X-ray			UV	S	IR			ESR			NMR	

$$-10 \quad -9 \quad -8 \quad -7 \quad -6 \quad -5 \quad -4 \quad -3 \quad -2 \quad -1 \quad 0 \quad 1$$
$$1\text{Å} \quad 1\text{ nm} \qquad\qquad 1\text{ μm} \qquad\qquad\qquad 1\text{ cm} \qquad 1\text{ m}$$
$$\log \lambda$$

The different methods can be divided into three parts:

(A) Electronic spectroscopy

(B) Vibrational spectroscopy

(C) Resonance spectroscopy

12.2 Electronic spectroscopy

These are caused by absorption of high energy photons, which can send electrons to higher energy levels. The electronic spectra are within the visible or ultraviolet regions of the electromagnetic spectrum. The spectral range extends at the wavelength scale used from 1.0 Å to 1.0 mm (=10,000 Å). The different methods used in this spectroscopy are:

(i)   Electron spectroscopy for chemical analysis (ESCA)

(ii) Ultraviolet spectroscopy (UV)

(iii) Visible spectroscopy (VIS)

ESCA spectrum informs about the binding energies or ionization energies of the electrons in the sample. ESCA is generally used to study polymer surfaces and also electronic structure of polymers. UV and VIS spectroscopy can be obtained from polymers containing aromatic, heteroaromatic, conjugated double, triple bonds or chromophores. So UV/VIS spectroscopy is a very important tool for studying the following parameters: (1) aromatic and heteroaromatic polymers, (2) photochemistry of polymers, (3) degradation of polymers and (4) polymeric radical ion complexes.

12.3 Vibrational spectroscopy

Lower energy photons cause changes in vibrational energy levels. The spectra thus obtained are referred to as vibrational spectroscopy. There are two methods present in vibrational spectroscopy, i.e. (1) Infrared (IR) spectroscopy and (2) Raman spectroscopy. The study of molecular structure by vibrational spectroscopy depends primarily upon the existence of vibratory motions of atoms within the molecule. Vibrational spectroscopy is one of the most widely used techniques for relatively rapid qualitative analysis and molecular structure determinations. This has resulted from the fact that molecular functional groups have characteristic 'group frequencies' which can be used for identification purposes.

The study of molecular structure by vibrational spectroscopy depends primarily upon the existence of vibratory motions of atoms within the molecule. The molecular motion that has the next higher energy level spacing after the rotation of molecules is the vibration of the atoms of the molecules with respect to one another. The study of the absorptions of radiation that result from transitions among the vibrational energy levels leads to further detailed insight into the nature of the molecules.

$$v = \frac{1}{2\pi c}\left|\frac{f}{\mu}\right|^{1/2} \tag{12.2}$$

$$\mu = m_x m_y / (m_x + m_y) \tag{12.3}$$

where v is the frequency in cycles/s; c, velocity of radiation; f, the force constant related to the strength of the bonds and μ, the reduced mass equivalent to $\mu = m_x m_y / (m_x + m_y)$; m_x and m_y are the masses of atoms of a bond. The vibratory motion depends upon f and μ, which in turn indicates the nature and arrangement of constituent atoms. It can be concluded that molecular functional groups have characteristic 'group frequencies', which can be used for analysis. The spectral range in case of vibrational spectroscopy extends from 1.0 μm to 1000 μm. There are two methods present in vibrational spectroscopy, i.e. (i) Infrared (IR) spectroscopy and (ii) Raman spectroscopy.

12.4 Infrared spectroscopy

The IR technique has been the predominant experimental method used to date because of the relative advances. IR spectrum is capable of giving indirect, but very valuable qualitative and quantitative structural information on polymer. The IR region is generally divided into three parts such as (a) Near IR or nIR, (b) Mid IR or mIR and (c) Far IR or fIR:

(a) Near Infrared region (nIR)

Near IR region extends from $v = 13,300 - 4000$ cm^{-1} (0.75 μ – 2.5 μ). The nIR spectra have been used to evaluate the structural order and quantitative composition of polymers. Near IR spectrum covers mostly stretching vibrations. Experimentally, the nIR is covered by many UV/visible spectrophotometers as nIR bonus using tungsten lamp as radiation source, quartz optics and a PbS detector.

(b) Mid Infrared region (mIR)

The major IR spectrum is the mIR spectroscopy. This region extends from 4000 to 200 cm^{-1} (2.5 μ–50 μ). The main application of mIR polymer spectroscopy is the analysis and identification of chemical structures and their changes. This application is followed at a higher level of sophistication, by studies of conformation, which also provide information about tacticity and crystallinity. Also the electronic structure influences the strength of many bonds and this causes shifts of maxima, which may help to elucidate the nature of bonding. Many mIR bands are due to the movement in the chain and crystal lattice (due to inter or intramolecular vibrations, known as photons). The schematic diagram of IR spectroscopy is given in Fig. 12.1.

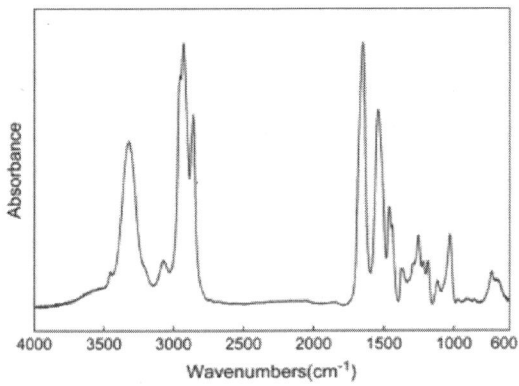

Figure 12.1 IR spectroscopy

(c) Far Infrared region (fIR)

FIR extends from $v = 200–10$ cm^{-1} (50 μ–1000 μ). Since there is no borderline between mIR and fIR, sometimes fIR extends up to 300–500 cm^{-1}. Absorption by polymers in the fIR is due to low frequency vibrations of heavy groups and/or weak bonds. The assignment of these low frequency spectra is based on the assumption of coupled torsional vibrations of molecular groups as a whole, neglecting internal motion of single groups. This spectrum also depends strongly on the degree of crystallinity.

The information available from IR is:

Structure	Dynamics
Chemical structure	Movements of chain segments of side groups
Tacticity	Movements in the crystal lattice
Conformations	Any complex formation and related phenomenon.
Crystallinity	
Electronic structure	

In addition to these information, specifically mIR is useful for analysis and identification of chemical structures as well as molecular structure of polymer by means of (1) structure and composition of copolymers, (2) molecular weight or molar mass, (3) chain branching in polymers, (4) presence of double bonds in saturated polymers, (5) tacticity, (6) presence of low molecular weight components after polymerization or at any stage of polymer processing, (7) crystallinity and (8) orientation. Absorption by polymers in the fIR region can be due to the following vibrations like (1) Valence vibrations of H–bonds: X–H ··· Y: mIR observed X–H or SX–H, modified by H bonds, (2) Intermolecular vibrations, (3) Torsional vibrations and (4) Collective lattice vibrations (phonons)

12.4.1 Fundamental vibrations

There are two kinds of fundamental vibrations for any molecules. Those are:

(a) Stretching vibration: The distance between two atoms increases or decreases but the atoms remains in the same bond axis.

(b) Bending or deformation vibration: The position of the atoms changes relative to the original bond axis.

The various stretching and bending vibrations of a bond occur at certain particular frequencies when infrared light of that same frequency is incident on the molecule, energy is absorbed and the molecules revert from the excited state to the original ground state, the absorbed energy is released as heat. The usual stretching and bending vibrations that can exist within a molecule are shown in Table 12.1.

Table 12.1   Stretching and bending vibrations

		Symbols	Examples
(A) Valence vibrations			

Stretching state	v	Sym v_s	Assym v_a
Breathing	br		
(B) Deformation (bending) vibrations			
Scissor vibrations	s		
Wagging	ω		
Rocking	r		
Twisting	t		
Torsion	τ		
Out of plane	Γ		
In plane bending	B		

The stretching vibrations or the stretching frequency (v in cm^{-1}) of a bond is related to the masses of the two atoms (m_x, m_y in grams), the velocity (c)

and force constants of the bonds (f in dynes/cm) as per Eq. (12.2). This means that if the bond present in the structure consists of very small proton, then stretching vibrations occur at much higher frequency, as in Table 12.2.

Table 12.2 Stretching vibrations of different groups

No.	Group	Frequency	
1	C–H	3.3 μ–3.5 μ	3100–2850 cm⁻¹
2	N–H	<2.8 μ–3.0 μ	3500–3300 cm⁻¹
3	O–H	2.7 μ–3.1 μ	3650–3200 cm⁻¹
4	O–D	3.8 μ	2630 cm⁻¹

f for single, double and triple bonds are approximately 5×10^5, 10×10^5, 15×10^5 dynes/cm, respectively. So the stretching vibrations are found to occur in the order of bond strengths, as shown in Table 12.3.

Table 12.3 Stretching vibrations of different bonds

No.	Bond	Quality	Frequency	
1	Triple bond	Strongest	4.4 μ–5.0 μ	2300–2000 cm⁻¹
2	Double bond	Stronger	5.3 μ–6.7 μ	1900–1500 cm⁻¹
3	Single bond	Strong	7.7 μ–12.5 μ	1300–800 cm⁻¹

Bending vibrations generally require less energy and occur at larger wavelength (lower wave number) than stretching vibrations.

12.4.2 Analysis of infrared spectrum

Empirical IR spectra are based on the existence of 'group vibrations'. The polymer spectra are more complicated than the monomer spectra due to

(a) Coincidence of transition energies, especially in symmetric molecules.

(b) Inefficient coupling between distant parts of the molecule. The vibrating groups are often separated by thousands of bands.

(c) Selection rules to be met regarding the change in dipole moment (IR).

The study of crystalline structure and/or molecular structure by IR spectroscopy depends primarily upon the existence of vibratory motions of atoms within the molecule. Depending upon the sensitiveness towards the structure, the vibratory motions are referred as (i) normal bands, (ii) crystalline bands, (iii) amorphous bands, (iv) special bands, (v) orientation sensitive bands, (vi) conformational bands, etc. The intensity of normal bands practically does not change upon transition from crystalline to amorphous state or solution. The intensity of crystalline bands is significantly higher in spectra of crystalline samples than in the spectra of amorphous or liquid samples. When band of this type can be observed in spectra of liquid and amorphous samples, they are shifted with respect to their position in the crystalline state. The intensity of amorphous bands is significantly higher in spectra of amorphous or liquid samples than in the spectra of crystalline samples. The special bands are sensitive to some structure of the fibre and these can also be reflected in Raman spectrum. Apart from these bands, there are some other bands, the intensities of which are sensitive to the particular structure like orientation sensitive bands, conformational bands. The spectral differences between these bands arise from differences in the distribution of the units in the respective structures.

Quantitative analysis of polymer systems can be done by proper processing system and it can be as follows:

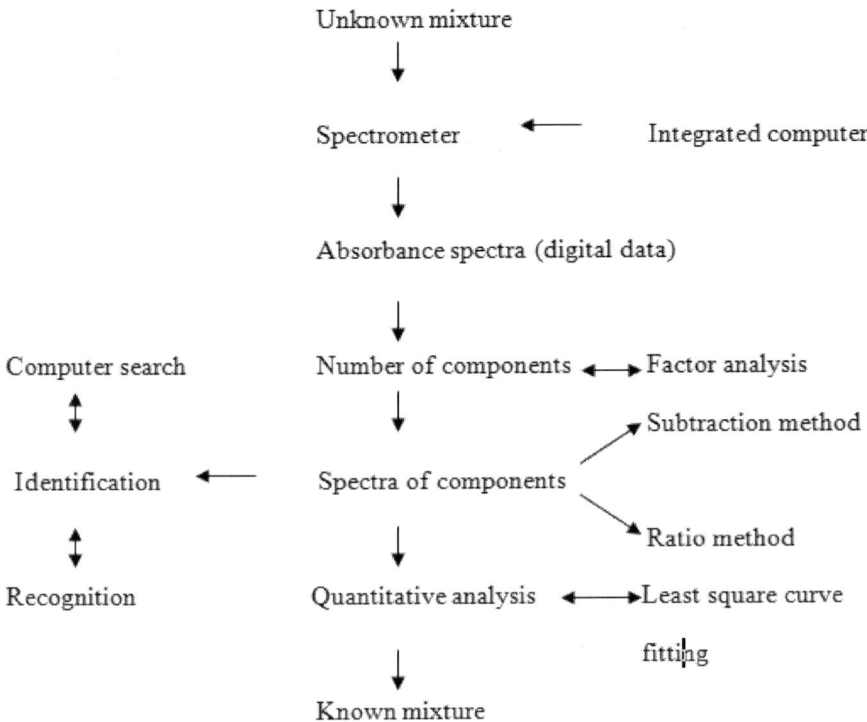

12.4.3 Measurement of crystallinity by IR spectroscopy

IR spectra can be used to measure degree of crystallinity. These are based on a common principle that the measured spectrum is the contribution from different sources. Because the infrared absorption spectra of the same polymer in the crystalline and amorphous states can differ for at least two reasons:

(1) Specific intermolecular interactions may exist in the crystalline polymer, which lead to sharpening or splitting of certain bonds.

(2) Some specific conformations may exist in one but not in the other phase, leading to bands, which are characteristic exclusively of either crystalline or amorphous material.

Measurement of crystallinity assumes a two phase model of the fibre structure. Based on this, the values of absorbance of the crystalline band as well as amorphous band can be as follows:

$$E_C = \alpha \cdot E_C^0 = \alpha \cdot \varepsilon_C \cdot d \tag{12.4}$$

$$E_A = \alpha \cdot E_A^0 = (1 - \alpha) \cdot \varepsilon_A \cdot d \tag{12.5}$$

where E_C is the measured absorbance of the crystalline band; E_A, measured absorbance of the amorphous band; α, degree of crystallinity; E_C^0, absorbance of 100% crystalline sample; E_A^0, absorbance of 100% amorphous band; ε_C, linear extinction co-efficient of 100% crystalline sample; ε_A, linear extinction co-efficient of 100% amorphous sample and d, thickness of the layer or concentration.

Ratio of E_C/E_A can be measured to cancel the value of d. The value of crystallinity from these equations will be as per Eq. (12.6), (12,7) or (12.8)

$$\alpha(\%) = \frac{E_C / E_A}{E_C / E_A + E_C^0 / E_A^0} \tag{12.6}$$

$$\alpha(\%) = \frac{1}{1 + (E_C^0 / E_A^0)(E_A / E_C)} \tag{12.7}$$

$$\alpha(\%) = \frac{1}{1 + (\varepsilon_C / \varepsilon_A)(E_A / E_C)} \tag{12.8}$$

ε_A can be measured from the amorphous phase in the melt and ε_C is the extrapolated values taken from density measurements used for calibration. Some structural sensitive bands present in PE and PA 6 are shown in Table 12.4.

Table 12.4   Sensitive bands

	Crystalline	Amorphous
PE	1894 cm^1	1305 cm^1

PA 6	γ	975 cm^{-1} (10.25 µm)	1120 cm^{-1} (8.93 µm) 1075 cm^{-1} (9.30 µm)
	α	957 cm^{-1} (10.45 µm)	
		930 cm^{-1} (10.75 µm)	
		833 cm^{-1} (12.0 µm)	

12.4.4 Tacticity and conformation

Tacticity is due to the pseudo-asymmetric carbon atoms. There is a possible link between configurations or tacticity and IR spectrum since regular configuration favours a confirmative order which influences IR absorption.

12.4.5 Chain folding

The vibrational bonds may provide information on chain folding or re-entry, molecular arrangement in the amorphous region like trans, avg, gauche, (cis) conformation, coiling, tie chain length distribution in the amorphous region. Also polarized IR radiation leads to absorption, which is a function of orientation of plane of polarization. This phenomenon is known as Dichroism or Dichroic Ratio. This happens for uniaxially oriented specimen occurs by stretching or doubly oriented specimen.

(i) Bands, the intensity of which practically does not change upon transition from crystalline to amorphous state or solution. This band corresponds to vibrations which are not appreciably affected by crystallinity.

(ii) Bands appearing only in crystalline nylon and bands the intensity of which is significantly higher in spectra of crystalline samples than in the spectra of amorphous or liquid samples. When bands of these types can be observed in spectra of liquid and amorphous samples, they are shifted with respect to their position in the crystalline state. These bands are sensitive to crystalline structure.

(iii) Bands appearing only in amorphous sample and bands the intensity of which is significantly higher in spectra of amorphous than in the spectra of crystalline samples.

12.4.6 Measurement techniques in IR spectroscopy

The measurement requires following steps:

(A) Sample preparation

(B) Wave number calibration

(C) Treatment of absorption intensity

(A)   Sample preparation

Sample preparation can be done by any of the following methods, which will be suitable for the sample. It should be noted that for structural characterization, the sample should not be heated or heat treated or undergoes any other treatment which will modify the structure. This will spoil the accuracy of the result. Some of the sample preparation methods are:

(1) Preparation of films either by casting from solution (thickness 20–80 μ), or by melting, or preparation from melt, rolling or heat pressing.

(2) Preparation of a paste using Nujol or hexachlorobutadiene as mulling agent.

(3) Preparation of a disc by means of KBr with or without addition of suitable solvent to promote swelling or partial dissolution. The disc should contain 200 mg KBr with 8 mg polymer powder.

(4) Arranging single filaments and fixed by a binding agent or by wetting with Nujol.

(5) Making the solution of the polymer.

(6) Preparation of microtone section.

(B)   Wave number calibration

The calibration can be done by means of polystyrene sample, as it gives sharp peaks in the range of 4000–650 cm^1.

(C)   Treatment of absorption intensity

The intensity of the IR absorption can be quantitatively expressed by means of Lambart–Beer law. This law is

$$I = I_o \exp(-k \cdot c \cdot t)$$

where I is the transmitted radiation intensity, I_o, incident radiation intensity, k, absorption co-efficient, c, concentration and t, thickness of the sample. Optical density of absorbance 'D' can be expressed from I_o and I by means of the following equation:

$$D = \log_{10}(I_o/I) = \log_{10} c \cdot \ln(I_o/I)$$
$$= 0.4343 \ln(I_o/I) = 0.4343 \cdot k \cdot c \cdot t$$

12.5 Raman spectroscopy

Raman spectroscopy is a scattering phenomenon, which changes the frequency of the incident light falling on a sample from v_o to another frequency v_v. The frequency difference $\Delta v = v_o - v_m$. The value of Δv may be either +ve or −ve and its magnitude is called 'Raman Spectroscopy' as per its discovery by Sir C. V. Raman in 1928.

Raman spectroscopy of polymers uses visible radiation in the spectral region between 400–600 nm in order to excite polymer vibrations. It is based on inelastic light scattering predicted theoretically by Smekal in 1923 and detected experimentally five years later by Raman. The inelastic light scattering is therefore called 'Smekal–Raman Effect', the spectroscopy based on this effect is generally called Raman spectroscopy. The advantages of Raman spectroscopy for structural determination are as follows:

1.   The intensity of the laser beam has enabled many fluorescence problems of coloured samples to be overcome.

2.   Allowed multiple pass techniques to increase the signal to noise ratio of weakly scattering sample.

3.   A combination of the intensity and nearly completely linear polarization of the beam has enabled accurate measurement of depolarization ratios, which indicate effects of neighbouring molecular structure and solved upon local group vibrations.

4.   It extends and confirms structural information from IR spectroscopy, since an adequate view of vibrational behaviour of a molecule is given by a combination of both types of spectra. Many weakly IR absorbing bands like C–C, C C, C C, and C–H display intense Raman scattering and vice versa.

5.   More distinctive fingerprint spectra are often obtained.

Raman spectroscopy is more connected with physical structures. In the linear polymers like PE, PP differences between crystalline, amorphous and solid/ molten state can be observed most easily. So conformations and crystallinity directly influence Raman spectra of polymers, whereas tacticity has an indirect influence by favouring certain conformations and crystal structures. In the amorphous and molten state, the intensity of reactions between molecules fluctuates randomly thus causing broadening of the spectra. In the crystalline phase, each molecule feels the same local environment so that the peaks are narrower because they are more homogeneous.

Raman spectroscopy can provide additional qualitative information about the introduction of branches or comonomers units into a polymer. As it greatly expands the range of melting, partial melting occurs continuously over a wide temperature range.

Raman spectroscopy has immense potentiality. The rapid growth in the use of this technique for fibre characterization is due to the emergence of new instrumentation. Raman spectroscopy is more connected with physical structures. In linear polymers like PE, PP differences between crystalline, amorphous and solid/molten state can be observed most easily. So conformations and crystallinity directly influence Raman spectra of polymers whereas tacticity has an indirect influence by favouring certain chain conformations and crystal structures. In the amorphous and molten state, the intensity of reactions between two molecules fluctuates randomly, thus causing broadening of the spectra. In the crystalline phase, each molecule feels the same local environment so that peaks are narrower because they are more homogeneous. Raman spectroscopy also can provide additional qualitative information about the introduction of branches or comonomer units into a polymer.

12.5.1 Advantages of Raman spectroscopy

The advantages of Raman spectroscopy are summarized as follows:

1. The intensity of the laser beam has enabled many fluorescence problems of coloured samples to be overcome.

2. Allowed multipass techniques to increase the signal-to-noise ratio of weakly scattering sample.

3. A combination of the intensity and nearly completely linear polarization of the beam has enabled accurate measurement of depolarization ratios, which indicate effects of neighbouring molecular structure and solvent upon local group vibrations.

4. It extends and confirms structural information from IR spectroscopy since an adequate view of vibrational behaviour of a molecule is given by a combination of both types of spectra. Many weakly IR absorbing bands display intense Raman scattering and vice versa, e.g. C–C, C=C, C≡C, C–H, etc.

5. More distinctive fingerprint spectra are often obtained.

12.5.2 Measurement of crystallinity by Raman spectroscopy

Raman spectroscopy method is based on the principle of resolving the spectrum into different contributions from the crystalline and disordered regions of the sample. The mass fractions in the different states of order follow from the weights of the respective components either directly or after calibration. The analysis of the spectrum is similar like that of IR spectroscopy. For example, stretching, bending and twisting of the bonds can give rise to Raman spectrum. However, Raman crystalline frequency bands are different from IR crystalline bands. The crystalline bands are mostly narrow bands, sometimes doublets. On the other hand, the amorphous or melt samples show broad scattering.

Raman spectroscopy has immense potentiality. The rapid growth in the use of this technique for fibre characterization is due to the emergence of new instrumentation. Raman spectroscopy is more connected with physical structures. At present, Raman spectroscopy has turned out to be an important tool in studying deformation characteristics of fibres, particularly high performance and highly oriented fibres on the molecular level.

12.6   Fourier transform infrared spectroscopy

Fourier Transform Infrared Spectroscopy (FTIR) is used instead of conventional IR spectroscopy. Typical modern FTIR equipment will provide (1) superb result, (2) engineered reliability, (3) high performance and (4) extensive application software. Computer controlled optical functions and automation features allow the user to easily optimize performance for particular application. FTIR provided better spectroscopic information with recent equipment. These may be (1) most advanced one for research, (2) a specialized instrument for specific application or (3) economical and powerful system for general works. All these are due to amazing development in computer systems for Fourier transform spectroscopy integrated with a high resolution colour graphics terminal, state-of-the-art storage devices, real time operating systems and extensive application software packages. The general features of FTIR include:

- instant display of spectra as they are acquired,

- effortlessly roll, zoom and expand any portion of a spectrum to view spectral details,

- overlay sample and reference data for direct visual comparison,

- simultaneously display as many spectra as derived for comparison,

- rapidly perform spectral subtractions and baseline corrections.

This means that once spectral data are collected, the FTIR equipment can then easily perform a variety of spectral manipulations. The special features of FTIR include:

1.   Fast scan facility, as well as wavelengths measured simultaneously,

2.   Availability of full IR energy for sending through sample,

3. The practical elimination of any moving parts,

4. High resolution,

5. Facility to work with complete spectral range from visible to far IR,

6. Interaction with computer.

Quantitative analysis of polymer systems can be done by data processing system in FTIR as per the following guidelines in addition to the data processing system mentioned for general IR spectroscopy.

The samples required for IR/FTIR spectroscopy should be very thin and it should be either in the form of films, paste, disc, and solution, microtomed or parallel arrangement of single filaments. For structural analysis, care should be taken in sample preparation with uniform cross-section. This will ensure high transmittance and better peak resolution for structural analysis. Details of structural analysis like conformational analysis, chain folding, crystallinity and types of crystallinity will be dealt in the next paper.

12.7 Resonance spectroscopic methods

Spectroscopy is the interaction between matter and electromagnetic radiation such that energy is absorbed or emitted according to the Bohr frequency condition. The group of spectroscopic methods that are based on the spin-resonance effect is due to true resonance absorption or emission of electromagnetic radiation under the additional action of external (machine parameters) and internal (molecular) magnetic fields. In spin-resonance spectroscopy, the interaction is responsible for absorption of the low energy, i.e. microwave and radio frequency are magnetic ones. The field of spin-resonance spectroscopy is classified into Electron Spin-Resonance (ESR) and Nuclear Magnetic Resonance (NMR).

12.7.1 Spin resonance

The groups of spectroscopic methods that are based on the spin-resonance effect are due to true resonance absorption or emission of electromagnetic radiation under the additional action of external or internal magnetic fields. Spectroscopy beyond fIR is non-resonance spectroscopy and it is due to slow movement of polymer chains in the microwave region. Spin is a movement, which is rotation about an axis of a body. This movement is characterized by an angular momentum and it is dependent upon the moment of inertia of the matter and angular velocity. Further, the angular momentum of any particle is related its quantum number. The various possible modes of motion like rotation, vibration, etc. that contribute to the total angular momentum may be treated independently and each assigned a separate quantum number. This quantum number specifies the angular momentum attributed to the particular mode of motion considered. The total energy of the particle may be considered as the sum of individual contributions due to various possible modes of motion.

The angular momentum (P) of any particle can be as per the following equation:

$$P = \theta \cdot \omega$$

$$P = (\xi) \cdot [R(R + 1)]^{1/2}$$

where θ is moment of inertia; ω, angular velocity; $\xi = h/2\varpi$; h, Planck's constant and R is a quantum number, either integral or half-integral. The various possible modes of motion (rotation, vibration, etc.) that contribute to the total angular momentum may be treated independently, and each assigned a separate quantum number. This quantum number specifies the angular momentum attributed to the particular mode of motion considered. Then the total energy of the particle may be considered as the sum of individual contributions due to various possible modes of motion.

12.7.2 Nuclear resonance

Nuclei of different elements and different isotopes of the same element differ in spin angular momentum. The features are: (i) Nuclei with an odd mass number have half-integral spin, e.g. H, C, N; (ii) Nuclei with an even mass number and an even charge number have zero spin, e.g. C, O and S and (iii) Nuclei with an even mass number but an odd charge number have integral spin, e.g. N. In general, NMR absorption lines of $I > 1/2$ nuclei are generally broad due to electric quadrupole moments. So it is not much of use as those for $I = 1/2$.

12.7.3 Nuclear magnetic moments

Any motion of a charged body has an associated magnetic field. Due to the motion of electrons along a conductor resulted electric current, which in turn produces such a field. This phenomenon also occurs on an atomic scale. So whenever electrons or nuclei possess angular momentum, there is a magnetic moment. When a substance is placed in a magnetic field, it becomes magnetized and modifies the field. The magnetization may be considered in terms of two contributions: (a) bulk or macroscopic effect, (b) local or microscopic effect. The bulk intensity of magnetization is proportional to the magnetic field. The micromagnet slightly modifies the magnetic field experienced by neighbouring protons. Owing to this, the polymers due to their rigidity and restricted molecular mobility show broadening of the peak absence of high resolution spectra. However, by using low viscosity solvents, the molecular movements will be rapid and it will exhibit high resolution NMR spectra.

The external magnetic field by the electronic environment can lead to the shielding of the nuclei and it ultimately gives rise to chemical shift. For a given compound, the appearance of the spectrum is governed by intramolecular chemical shift differences, i.e. differences in resonance frequencies for different nuclei of the same molecule. For this reason, the standard solution and the standard spectra should (i) contain only a single type of magnetic nucleus with a single type of environment so that NMR spectra consists of one sharp line, (ii) contain many nuclei of that type and environment for an intense signal. Further, the standard should be chemically inert so that its resonance position is not affected by the sample and the resonance should occur in a region of the spectrum away from the signals given by most common chemicals.

Nuclear magnetic moments can also lead to spin–spin coupling. These occur because of splitting of the bands due to each nucleus. For example, if there are two nuclei with non-zero spin in the molecule, then it may be possible that the resonance of each spin is split into two lines of equal intensity. The line separation is equal. The nature of splitting is dependent upon the molecules and their masses. This interaction is generally referred as spin–spin coupling and its magnitude is generally determined by spin–spin coupling constant. In turn, the coupling constant is dependent upon chemical environment and so it is of great use for structural determination.

12.8   Nuclear magnetic resonance spectroscopy (NMR)

NMR spectroscopy is related to the transitions between nuclear spin energy levels. It involves the magnetic energy of nuclei when they are placed in a magnetic field and the transitions occur in the radio wave region of the spectrum.

The transitions between nuclear spin energy levels give rise to the phenomenon of NMR. It involves the magnetic energy of nuclei when they are placed in a magnetic field, and the transitions occur in the radii wave region of the spectrum. NMR is observed in strong magnetic fields at excitation frequencies of 20–360 MHz, which corresponds to λ of 0.8 mm–15 mm. However, the interaction between the magnetic particles and the external magnetic field is much weaker in NMR because of weakness of nuclear magnetic moments. So the usefulness of NMR is restricted to nuclei with a nuclear spin not equal to zero and thus with a finite nuclear moment. The nuclear spin is the total spin of protons and spin of neutrons. Some important nuclei with magnetic moment are ^{1}H, ^{13}C, ^{14}N and ^{19}F. The most important nucleus in polymer spectroscopy is ^{13}C since nearly all polymers have carbon atom in the main chain. Nuclei of different elements and different isotopes of the same element differ in spin angular momentum. The features are given in the below table.

(a) Nuclei with an odd mass number have half-integral spin, e.g. ^{1}H, ^{13}C, ^{15}N.

(b) Nuclei with an even mass number and an even charge number have zero spin, e.g. ^{12}C, ^{16}O, ^{32}S.

(c) Nuclei with an even mass number but an odd charge number have an integral spin, e.g. ^{14}N.

Mass number	Atomic number	Spin number	Example
Odd	Odd or even	1/2, 3/2, 5/2	^{1}H, ^{13}C, ^{15}N
Even	Even	0	^{12}C, ^{16}O, ^{32}S
Even	Odd	1, 2, 3	^{14}N

In general, NMR absorption lines of $I > \frac{1}{2}$ nuclei are generally broad due to electric quadrupole moments. So it is not much of use as those for $I > \frac{1}{2}$.

^{1}H NMR

This is used in solutions of low viscosity. It will give small line width. The information available from ^{1}H NMR is as follows:

1. Identification of polymers by comparison with reference spectrum like that of IR spectroscopy.

2. Chemical shift and number of protons for unknown molar mass compounds.

3. Spin–spin coupling for unknown molar mass compounds.

4. Configuration (tacticity) analysis by chemical shifting and by spin–spin coupling.

^{13}C NMR

The most important nucleus in polymer spectroscopy is ^{13}C because of the following reasons:

1. Nearly all polymers consist of carbon atoms in the main chain as well as in side groups.

2. The range of chemical shift is very large.

3. Spin–spin coupling between neighbouring (^{13}C–^{13}C) atoms is very weak.

This system is advantageous than that of ^{1}H NMR because of the reasons that (1) large number of chemical shift enabling the resolution of chemically similar structures and (2) minimum dipolar broadening due to higher segmental mobility. ^{13}C NMR can provide the following information:

1. Identification of specific polymer,

2. Identification of specific groups in a polymer,

3. Branching of macromolecule,

4. Analysis of sequence in copolymers,

5. Distinction between block and random copolymers,

6. Analysis of tacticity.

Structure	Dynamics
Chemical structure	Movements of chain segments or side groups
Tacticity	
Conformations	

NMR spectroscopy is employed for determination of the microstructure, chemical composition, and chemical inhomogeneity branching in polymers. Measurement of the mobility of single polymer group can also be carried out. Measurement of different proton mobility in amorphous and crystalline regions provides information regarding crystallinity of the polymer, the chain conformation or mechanism of molecular motions at different temperatures. NMR spectroscopy has been the only method, which provided direct evidence for the occurrence of a third intermediate state of order. It is likely that the regions of 'restricted mobility' reflected in the intermediate component of the NMR signal and the anisotropic disordered regions showing up the Raman spectrum are largely identical.

12.8.1 Nuclear magnetic resonance of polymers

It is always important to select a nucleus for the determination of polymer structure by nuclear magnetic resonance. There are two nuclei available for the structural determination. Those are: H and C. C is more preferable because of the following reasons: (i) nearly all polymers show carbon (C) atoms in their main chain, (ii) larger range of chemical shift enabling the resolution of chemically similar structures, (iii) minimum dipolar broadening and (iv) weaker spin–spin coupling between neighbouring atoms. The data available from NMR spectroscopy are: (1) Identification of specific polymers, (2) identification of groups in a polymer, (3) Branching of macromolecules, (4) Analysis of sequences in copolymers, (5) Distinction between block and random copolymers, (6) Analysis of tacticity, (7) Analysis of conformation.

12.9   Electron spin-resonance spectroscopy (ESR)

The structure of the periodic table can be explained on the basis of four electronic quantum numbers. These are n, l, m_i, m_s, where m_s is the directional spin quantum number. So electrons also possess angular momentum. It could be attributed to a spinning motion, i.e. ESR. The basis of ESR spectroscopy is absorption of electromagnetic radiation by unpaired electrons in a magnetic field. Unpaired electrons are characterized by an angular momentum, the spin, which is connected with a magnetic moment. The magnetic moment is influenced by external magnetic field. The resonance frequency of ESR depends not only on molecular properties but also on external magnetic field. Since polymers are almost exclusively diamagnetic, ESR studies are restricted to the detection of usually small concentrations of polymer molecules containing unpaired electrons or polymer radicals. ESR gives precise information about radicals formed by radiation, polymerization, photolysis or mechanical stress like fracture and deformation process in fibres. Measurement of radical formation sometimes provides information regarding the number of taut tie molecules extending between crystalline regions. Otherwise ESR gives no direct information about the polymers or fibres. Sp the main fields of polymer related applications of ESR are:

- ➤ Radical induced polymerization
- ➤ Mechanical degradation and fracture of polymers
- ➤ Radiation and light induced radical formation
- ➤ Spin label and spin probe techniques
- ➤ Electron spin polymers
- ➤ Conductive polymers
- ➤ Triplet stated and biradicals

Direct evidence of radical mechanism of polymerization is obtained by detection of radical intermediates, and ESR can do it.

Structure	Dynamics
Chemical structure	Movements of chains and side groups
Electrical structure	Complex formation

The basis of ESR spectroscopy is absorption of electromagnetic radiation by unpaired electrons in a magnetic field. Since polymers are almost exclusively diamagnetic, ESR studies are restricted to the detection of usually small concentrations of polymer molecules containing unpaired electrons or polymer radicals. So ESR gives precise information about radicals formed by radiation, polymerization, photolysis or mechanical stress like fracture and deformation process in fibres. Measurement of radical formation sometimes provides information regarding the number of taut tie molecules extending between crystalline regions. Otherwise ESR gives no direct information about the polymers or fibres.

Further readings

1. D. Campbell, R. A. Pethrick, J. R. White, *Polymer Characterization Physical Techniques*. Chapman and Hall, 1989.

2. *Encyclopedia of Polymer Science and Engineering*, Wiley, New York, 1986.

3. R.S. Stein, *Newer Methods of Polymer Characterization*, R. Ke (Ed.), Interscience Publishers, New York, 1964, p. 185.

4. J. L. Koenig, *Spectroscopy of Polymers*, Elsevier, 1999.